AN ALGEBRAIC INTRODUCTION TO COMPLEX PROJECTIVE GEOMETRY. 1

An Algebraic Introduction to Complex Projective Geometry

1. Commutative algebra

Christian Peskine
Professor at University Paris VI, Pierre et Marie Curie

CAMBRIDGE
UNIVERSITY PRESS

CAMBRIDGE UNIVERSITY PRESS
Cambridge, New York, Melbourne, Madrid, Cape Town, Singapore, São Paulo, Delhi

Cambridge University Press
The Edinburgh Building, Cambridge CB2 8RU, UK

Published in the United States of America by Cambridge University Press, New York

www.cambridge.org
Information on this title: www.cambridge.org/9780521108478

First published 1996
This digitally printed version 2009

A catalogue record for this publication is available from the British Library

ISBN 978-0-521-48072-7 hardback
ISBN 978-0-521-10847-8 paperback

Contents

Contents

16 Cartier divisors **211**
16.1 Cartier divisors 211
16.2 The Picard group 217
16.3 Exercises 220

Subject index *225*
Symbols index *229*

Introduction

Commutative algebra is the theory of commutative rings and their modules. Although it is an interesting theory in itself, it is generally seen as a tool for geometry and number theory. This is my point of view. In this book I try to organize and present a cohesive set of methods in commutative algebra, for use in geometry. As indicated in the title, I maintain throughout the text a view towards complex projective geometry.

In many recent algebraic geometry books, commutative algebra is often treated as a poor relation. One occasionally refers to it, but only reluctantly. It also suffers from having attracted too much attention thirty years ago. One or several texts are usually recommended: the "Introduction to Commutative Algebra" by Atiyah and Macdonald is a classic for beginners and Matsumura's "Commutative rings" is better adapted for more advanced students. Both these books are excellent and most readers think that there is no need for any other. Today's students seldom consult Bourbaki's books on commutative algebra or the algebra part in the E.G.A. of Grothendieck and Dieudonné.

With this book, I want to prepare systematically the ground for an algebraic introduction to complex projective geometry. It is intended to be read by undergradute students who have had a course in linear and multilinear algebra and know a bit about groups. They may have heard about commutative rings before, but apart from $\mathbb{Z}$ and polynomial rings in one variable with coefficients in $\mathbb{R}$ or $\mathbb{C}$, they have essentially worked with fields. I had to develop quite a lot of language new to them, but I have been careful to articulate all chapters around at least one important theorem. Furthermore I have tried to stimulate readers, whenever their attention may be drifting away, by presenting an example, or by giving them an exercise to solve.

In the first eight chapters, the general theory of rings and modules is developed. I put as much emphasis on modules as on rings; in modern algebraic geometry, sheaves and bundles play as important a role as varieties. I had to decide on the amount of homological algebra that should be included and on the form it should take. This is difficult since the border between commutative and homological algebras is not well-defined. I made several conventional choices. For example, I did not elaborate immediately on the homological nature of length. But quite early on, when studying dualizing modules on Artinian rings in chapter six, I used non-elementary homological methods. In chapter seven, I have been particularly careful on rings and modules of fractions, hoping to prepare readers for working with sheaves.

I wanted this book to be self-contained. Consequently, the basic Galois theory had to be included. I slipped it in at the end of this first part, in chapter nine, just after the study of integral ring extensions.

Now, our favourite ring is $\mathbb{C}[X_1, \ldots, X_n]$, the polynomial ring in several variables with coefficients in the field of complex numbers. We can derive many rings from this one by natural algebraic procedures, even though our purposes are geometric. Quotient rings of $\mathbb{C}[X_1, \ldots, X_n]$, in other words, finitely generated $\mathbb{C}$-algebras, and their fraction rings are the basic objects from chapter ten to chapter thirteen. Noether's normalisation lemma and Hilbert's Nullstellensatz, two splendid theorems of commutative algebra, concern these rings and are at the heart of algebraic geometry. With these results in view, I discuss the notion of dimension and move heartily towards geometry by introducing affine complex schemes and their morphisms. I can then present and prove two other important geometric results, a local version of Zariski's main theorem and Chevalley's semi-continuity theorem.

From chapter fourteen on, I have tried to provide a solid background for modern intersection theory by presenting a detailed study of Weil and Cartier divisors.

In order to keep this book short, I have had to make many painful choices. Several of the chapters that I have deleted from this text will appear in a second book, intended for graduate students, and devoted to homological algebra and complex projective geometry.

I have been careless with the historical background. But I have been careful in developing the material slowly, at least initially, though it does become progressively more difficult as the text proceeds. When necessary, examples and exercises are included within chapters. I allow myself to refer to some of those. All chapters are followed by a series of exercises. Many are easy and a few are more intricate; readers will have to make their own evaluation!

I would like to thank the many students, undergraduates or graduates, whom I taught, or with whom I discussed algebra, at the universities of Strasbourg, Oslo and Paris VI. I have tried to attract each of them to algebra and algebraic geometry. Special thanks are due to Benedicte Basili who wrote a first set of notes from my graduate course in algebraic geometry and introduced me to LaTeX.

My wife Vivi has been tremendously patient while I was writing this first book. I hope very much that some readers will think that she did well in being so.

1

Rings, homomorphisms, ideals

Our reader does not have to be familiar with commutative rings but should know their definition. Our rings always have an identity element 1. When necessary we write 1_A for the identity element of A. The zero ring $A = \{0\}$ is the only ring such that $1_A = 0$. In the first two sections we recall the really basic facts about ideals and homomorphisms (one of the reasons for doing so is because we need to agree on notation). From section 3 on, we begin to think about algebraic geometry. Prime and maximal ideals are the heart of the matter. Zariski topology, the radicals and comaximal ideals are henceforth treated. Our last section is a first approach to unique factorization domains (UFDs) (the proof of an essential theorem is postponed to chapter 7).

Examples 1.1

1. $\mathbb{Z}, \mathbb{Q}, \mathbb{R}$ and $\mathbb{C}$ are rings. Each of them is a subring of the next.

2. A commutative field K, with identity element, is a non-zero ring such that $K \setminus \{0\}$ is a multiplicative group.

3. The polynomial ring $K[X_1, ..., X_n]$ is a ring of which $K[X_1, ..., X_{n-1}]$ is a subring.

4. If A is a ring then $A[X_1, ..., X_n]$ is a ring of which $A[X_1, ..., X_{n-1}]$ is a subring.

5. If A and B are two rings, the product $A \times B$ has a natural ring stucture

$$(a, b) + (a', b') = (a + a', b + b') \quad \text{and} \quad (a, b)(a', b') = (aa', bb').$$

Exercises 1.2

1. If K and K' are two fields, verify that the ring $K \times K'$ is not a field.

2. Let p be a prime number. Denote by $\mathbb{Z}_{(p)}$ the subset of $\mathbb{Q}$ consisting of all n/m such that $m \notin p\mathbb{Z}$. Verify that $\mathbb{Z}_{(p)}$ is a subring of $\mathbb{Q}$.

3. Let $x = (x_1, ..., x_n) \in K^n$ be a point. Verify that the set of all P/Q, with $P, Q \in K[X_1, ..., X_n]$, and $Q(x_1, ..., x_n) \neq 0$, is a subring of the field $K(X_1, ..., X_n)$.

Definition 1.3 *Let A and B be rings. A (ring) homomorphism $f : A \to B$ is a set application such that for all $x, y \in A$.*

$$f(1_A) = 1_B, \quad f(x+y) = f(x) + f(y) \quad \text{and} \quad f(xy) = f(x)f(y).$$

An A-algebra is a ring B with a ring homomorphism $f : A \to B$.

The composition of two composable homomorphisms is clearly a homomorphism.

1.1   Ideals. Quotient rings

Proposition 1.4 *The kernel $\ker f = f^{-1}(0)$, of a ring homomorphism $f : A \to B$ is a subgroup of A such that*

$$(a \in \ker f \quad \text{and} \quad x \in A) \quad \Rightarrow \quad ax \in \ker f.$$

This is obvious.

Definition 1.5 *A subgroup $\mathcal{I}$ of a ring A is an ideal of A if*

$$(a \in \mathcal{I} \quad \text{and} \quad x \in A) \quad \Rightarrow \quad ax \in \mathcal{I}.$$

If $\mathcal{I} \neq A$, we say that $\mathcal{I}$ is a proper ideal.

Examples 1.6

1. The kernel of a ring homomorphism $f : A \to B$ is an ideal of A.
2. If $n \in \mathbb{Z}$, then $n\mathbb{Z}$ is an ideal of $\mathbb{Z}$.
3. Let $a \in A$. The set aA of all multiples of a is an ideal of A.
4. More generally, let a_i, with $i \in E$, be elements in A. The set of all linear combinations, with coefficients in A, of the elements a_i is an ideal. We say that the elements a_i, $i \in E$, form a system of generators of (or generate) this ideal, which we often denote by $((a_i)_{i \in E})$.

As an obvious but important remark we note that all ideals contain 0.

Exercise 1.7 Show that a ring A is a field if and only if (0) is the unique proper ideal of A.

Theorem 1.8 *Let $\mathcal{I}$ be an ideal of A and $A/\mathcal{I}$ the quotient group of equivalence classes for the relation $a \sim b \iff a - b \in \mathcal{I}$. Then $A/\mathcal{I}$ has a ring structure such that the class map $\mathrm{cl} : A \to A/\mathcal{I}$ is a ring homomorphism (obviously surjective) with kernel $\mathcal{I}$.*

Proof It is obvious that $\mathrm{cl}(a + b)$ and $\mathrm{cl}(ab)$ only depend on $\mathrm{cl}(a)$ and $\mathrm{cl}(b)$. Defining then

$$\mathrm{cl}(a) + \mathrm{cl}(b) = \mathrm{cl}(a + b) \quad \text{and} \quad \mathrm{cl}(a)\mathrm{cl}(b) = \mathrm{cl}(ab),$$

the theorem is proved.   □

Definition 1.9 *The ring $A/\mathcal{I}$ is the quotient ring of A by the ideal $\mathcal{I}$.*

Example 1.10 If $n \in \mathbb{Z}$, the quotient ring (with $|n|$ elements) $\mathbb{Z}/n\mathbb{Z}$ is well known.

Exercise 1.11 Let K be a field. Show that the composition homomorphism

$$K[Y] \xrightarrow{i} K[X,Y] \xrightarrow{\mathrm{cl}} K[X,Y]/XK[X,Y]$$

is an isomorphism (i is the natural inclusion and cl the class application).

Let B be a quotient of $A[X_1, ..., X_n]$. The obvious composition homomorphism $A \to B$ gives to B the structure of an A-algebra. We note that any element in B is a combination, with coefficients in A, of products of the elements $\mathrm{cl}(X_1), ..., \mathrm{cl}(X_n) \in B$. These elements generate B as an A-algebra.

Definition 1.12 *A quotient ring B of a polynomial ring $A[X_1, ..., X_n]$ over a ring A is called an A-algebra of finite type or a finitely generated A-algebra.*
Putting $x_i = \mathrm{cl}(X_i) \in B$, we denote B by $A[x_1, ..., x_n]$ and we say that $x_1, \ldots, x_n$ generate B as an A-algebra.

Clearly a quotient of an A-algebra of finite type is an A-algebra of finite type.

Theorem 1.13 *(The factorization theorem)*
Let $f : A \to B$ be a ring homomorphism. There exists a unique injective ring homomorphism $g : A/\ker f \to B$ such that the following diagram is commutative:

$$
\begin{array}{ccc}
A & \xrightarrow{f} & B \\
\downarrow{\mathrm{cl}} & \nearrow{g} & \\
A/\ker f & &
\end{array}
$$

Furthermore f is surjective if and only if g is an isomorphism.

Proof One verifies first that $f(a)$ only depends on $\mathrm{cl}(a) \in A/\ker f$. If we put then $g(\mathrm{cl}(a)) = f(a)$, it is clear that g is a well-defined injective homomorphism. The rest of the theorem follows easily.   □

The proof of the following proposition is straightforward and left to the reader.

Proposition 1.14 *Let A be a ring and $\mathcal{I}$ an ideal of A.*

(i) *If $\mathcal{J}$ is an ideal of $A/\mathcal{I}$, then $\mathrm{cl}^{-1}(\mathcal{J})$ is an ideal of A containing $\mathcal{I}$.*

(ii) *If $\mathcal{I}'$ is an ideal of A containing $\mathcal{I}$, then $\mathrm{cl}(\mathcal{I}')$ is an ideal of $A/\mathcal{I}$ (denoted $\mathcal{I}'/\mathcal{I}$).*

(iii) *One has $\mathrm{cl}^{-1}(\mathrm{cl}(\mathcal{I}')) = \mathcal{I}'$ and $\mathrm{cl}(\mathrm{cl}^{-1}(\mathcal{J})) = \mathcal{J}$. This bijection between the set of ideals of A containing $\mathcal{I}$ and the set of ideals of $A/\mathcal{I}$ respects inclusion.*

Note that this can be partly deduced from the next result which is useful by itself.

Proposition 1.15 *If $\mathcal{J}$ is an ideal of $A/\mathcal{I}$, then $\mathrm{cl}^{-1}(\mathcal{J})$ is the kernel of the composed ring homomorphism*

$$A \to A/\mathcal{I} \to (A/\mathcal{I})/\mathcal{J}.$$

This homomorphism factorizes through an isomorphism

$$A/\mathrm{cl}^{-1}(\mathcal{J}) \simeq (A/\mathcal{I})/\mathcal{J}.$$

This description of $\mathrm{cl}^{-1}(\mathcal{J})$ needs no comment. The factorization is a consequence of the factorization theorem.

Definition 1.16

(i) *An ideal generated by a finite number of elements is of finite type (or finitely generated).*

(ii) *An ideal generated by one element is principal.*

If $a_1, ..., a_n$ generate the ideal $\mathcal{J}$, we write $\mathcal{J} = (a_1, ..., a_n)$.

Theorem 1.17

(i) *All ideals in $\mathbb{Z}$ are principal.*

(ii) *If K is a field, all ideals in the polynomial ring (in one variable) $K[X]$ are principal.*

Proof Showing that a non-zero ideal $\mathcal{I}$ of $\mathbb{Z}$ is generated by the smallest positive integer of $\mathcal{I}$ is straightforward.

Following the same principle, let $\mathcal{I}$ be a non-zero ideal of $K[X]$. If $P \in \mathcal{I}$ is a non-zero polynomial such that $\deg(P) \le \deg(Q)$ for all non-zero polynomials $Q \in \mathcal{I}$, showing that $\mathcal{I} = PK[X]$ is also straightforward. $\qquad\square$

Definition 1.18 *Let A be a ring.*

(i) *If $a \in A$ is invertible, in other words if there exists $b \in A$ such that $ab = 1$, then a is a unit of A. One writes $b = a^{-1}$ and says that b is the inverse of a.*

(ii) *If $a \in A$ and $b \in B$ are elements such that $ab = 0$ and $b = 0$, we say that a is a zero divider.*

(iii) *If $a \in A$ is such that there exists an integer $n > 0$ such that $a^n = 0$, then a is nilpotent.*

Examples 1.19

1. The only units in $\mathbb{Z}$ are 1 and -1.

2. The units of $K[X]$ are the non-zero constants.

3. An element $\operatorname{cl}(m) \in \mathbb{Z}/n\mathbb{Z}$ is a unit if and only if m and n are relatively prime.

4. The ring $\mathbb{Z}/n\mathbb{Z}$ has no zero divisors if and only if n is prime.

5. The ring $\mathbb{Z}/n\mathbb{Z}$ has a non-zero nilpotent element if and only if n has a quadratic factor.

6. If $\mathcal{I} = (X^2 + Y^2, XY) \subset K[X,Y]$, then $\operatorname{cl}(X + Y), \operatorname{cl}(X)$ and $\operatorname{cl}(Y)$ are nilpotent elements of $K[X,Y]/\mathcal{I}$.

Definition 1.20

(i) *A non-zero ring without zero divisors is called a domain.*

(ii) *A non-zero ring without non-zero nilpotent elements is called a reduced ring.*

Definition 1.21 *A domain which is not a field and such that all its ideals are principal is a principal ideal ring.*

Hence our Theorem 1.17 can be stated in the following way.

Theorem 1.22 *The domains $\mathbb{Z}$ and $K[X]$ are principal ideal rings.*

1.2   Operations on ideals

Exercise 1.23 If $\mathcal{I}$ and $\mathcal{J}$ are ideals of a ring A, then $\mathcal{I} \cap \mathcal{J}$ is an ideal of A.

Note that if $\mathcal{I}$ and $\mathcal{J}$ are ideals of a ring A, then $\mathcal{I} \cup \mathcal{J}$ is not always an ideal of A.

Definition 1.24 *If $\mathcal{I}$ and $\mathcal{J}$ are ideals of a ring A, then*

$$\mathcal{I} + \mathcal{J} = \{a + b, \quad a \in \mathcal{I}, \; b \in \mathcal{J}\}.$$

Note that $\mathcal{I} + \mathcal{J}$ is an ideal of A, obviously the smallest ideal containing $\mathcal{I}$ and $\mathcal{J}$.

Definition 1.25 *Let $\mathcal{I}_s$ be a family of ideals of A. We denote by $\sum_s \mathcal{I}_s$ the set formed by all finite sums $\sum_s a_s$, with $a_s \in \mathcal{I}_s$.*

We note once more that $\sum_s \mathcal{I}_s$ is an ideal of A, the smallest ideal containing $\mathcal{I}_s$ for all s.

Definition 1.26 *If $\mathcal{I}$ and $\mathcal{J}$ are ideals of A, the product $\mathcal{I}\mathcal{J}$ denotes the ideal generated by all ab with $a \in \mathcal{I}$ and $b \in \mathcal{J}$.*

Definition 1.27 *If $\mathcal{I}$ is an ideal of A and P a subset of A, we denote by $\mathcal{I} : P$ the set of all $a \in A$ such that $ax \in \mathcal{I}$ for all $x \in P$.*

If $P \not\subset I$, then $\mathcal{I} : P$ is a proper ideal of A. If $P \subset \mathcal{I}$, then $\mathcal{I} : P = A$.

Exercise 1.28 If $\mathcal{I}_i$, $i = 1, ..., n$, are ideals of A and P a subset of A, show that

$$\left(\bigcap_i \mathcal{I}_i\right) : P = \bigcap_i (\mathcal{I}_i : P).$$

Proposition 1.29 *If $f : A \to B$ is a ring homomorphism and $\mathcal{J}$ an ideal of B, then $f^{-1}(\mathcal{J})$ is an ideal of A (the contraction of $\mathcal{J}$ by f).*

Proof The kernel of the composition homomorphism $A \to B \to B/\mathcal{J}$ is $f^{-1}(\mathcal{J})$.   □

Definition 1.30 *If $f : A \to B$ is a ring homomorphism and if $\mathcal{I}$ is an ideal of A, we denote by $f(\mathcal{I})B$ the ideal of B generated by the elements of $f(\mathcal{I})$.*

In other words, $f(\mathcal{I})B$ is the set consisting of all sums of elements of the form $f(a)b$ with $a \in \mathcal{I}$ and $b \in B$.

1.3 Prime ideals and maximal ideals

Definition 1.31 *An ideal $\mathcal{I}$ of A is prime if the quotient ring $A/\mathcal{I}$ is a domain.*

We note that a prime ideal has to be proper. The following result is obvious.

Proposition 1.32 *An ideal $\mathcal{I}$ is prime if and only if*

$$ab \in \mathcal{I} \quad \text{and} \quad a \notin \mathcal{I} \implies b \in \mathcal{I}.$$

Definition 1.33 *An ideal $\mathcal{I}$ of A is maximal if the quotient ring $A/\mathcal{I}$ is a field.*

Clearly, a maximal ideal is a prime ideal. The terminology is a bit unpleasant: obviously A is an absolute maximum in the set of all ideals, ordered by the inclusion. But the word maximal is justified by the following result:

Proposition 1.34 *An ideal $\mathcal{I}$ is maximal if and only if $\mathcal{I}$ is a maximal element of the set of all proper ideals, ordered by the inclusion.*

Proof A field is a ring whose only ideal is (0). Our proposition is an immediate consequence of Proposition 1.14. □

Exercises 1.35

1. Let k be a field and $(a_1, ..., a_n) \in k^n$. Show that the set of all polynomials $P \in k[X_1, ..., X_n]$, such that $P(a_1, ..., a_n) = 0$, is a maximal ideal of $k[X_1, ..., X_n]$ generated by $X_1 - a_1, ..., X_n - a_n$.

2. Show that all non-zero prime ideals of a principal ideal ring are maximal.

Proposition 1.36 *Let $\mathcal{P}$ be a prime ideal. If $\mathcal{I}_i$, with $i = 1, ..., n$, are ideals such that $\bigcap_1^n \mathcal{I}_i \subset \mathcal{P}$, there exists l such that $\mathcal{I}_l \subset \mathcal{P}$.*

Proof Assume not; then there exist $a_i \in \mathcal{I}_i$ and $a_i \notin \mathcal{P}$ for $i = 1, ..., n$. Since $\mathcal{P}$ is a prime ideal, this implies $\prod_1^n a_i \notin \mathcal{P}$. But $\prod_1^n a_i \in \bigcap_1^n \mathcal{I}_i$; this is a contradiction. □

Theorem 1.37 *(Avoiding lemma)*
Let $\mathcal{I}_1, .., \mathcal{I}_n$ be ideals of A such that at most two are not prime. If $\mathcal{J}$ is an ideal of A such that $\mathcal{J} \not\subset \mathcal{I}_m$ for $m = 1, ..., n$, then $\mathcal{J} \not\subset \bigcup_1^n \mathcal{I}_m$.

Proof We use induction on n. The result is clear for $n = 1$.

If $n > 1$, by the induction hypothesis, there exists, for all i, an element $a_i \in \mathcal{J}$, such that $a_i \notin \bigcup_{m \neq i} \mathcal{I}_m$. We can obviously assume $a_i \in \mathcal{I}_i$ for all i. Assume that $\mathcal{I}_1$ is prime if $n > 2$ and put $a = a_1 + \prod_{i>1} a_i$.

For $i > 1$ we have $a_1 \notin \mathcal{I}_i$ and $\prod_{i>1} a_i \in \mathcal{I}_i$. This shows $a \notin \mathcal{I}_i$ for $i > 1$. Since $\prod_{i>1} a_i \notin \mathcal{I}_1$ and $a_1 \in \mathcal{I}_1$, we have $a \notin \mathcal{I}_1$ and we are done. $\qquad\square$

Note that we have proved in fact the following more general useful result.

Theorem 1.38 *Let $\mathcal{I}_1, .., \mathcal{I}_n$ be ideals of A such that at most two are not prime. If E is a subset of A, stable for addition and multiplication, such that $E \not\subset \mathcal{I}_m$ for $m = 1, ..., n$, then $E \not\subset \bigcup_1^n \mathcal{I}_m$.*

Proposition 1.39 *Let $\mathcal{I}$ be an ideal of A and let $\mathcal{J}$ be an ideal of A containing $\mathcal{I}$. Then $\mathcal{J}$ is a prime (resp. maximal) ideal of A if and only if $\mathcal{J}/\mathcal{I}$ is a prime (resp. maximal) ideal of $A/\mathcal{I}$.*

Proof The proposition is an immediate consequence of the ring isomorphism

$$A/\mathcal{J} \simeq (A/\mathcal{I})/(\mathcal{J}/\mathcal{I}).$$

$\qquad\square$

Theorem 1.40 *A ring $A \neq 0$ has a maximal ideal.*

Let us first recall Zorn's lemma (or axiom):

Let E be a non-empty ordered set. If all totally ordered subsets of E are bounded above, E contains a maximal element.

Proof Consider $\mathcal{I}_i$, a totally ordered set of proper ideals of A. Define $\mathcal{I} = \bigcup_i \mathcal{I}_i$. We show that $\mathcal{I}$ is a proper ideal of A (obviously an upper bound for our totally ordered set).

If $a, b \in \mathcal{I}$ and $c \in A$, there exists $i \in E$ such that $a, b \in \mathcal{I}_i$. This implies $a + b \in \mathcal{I}_i \subset \mathcal{I}$ and $ac \in \mathcal{I}_i \subset \mathcal{I}$. Furthermore, since $1_A \notin \mathcal{I}_i$ for all i, it is clear that $1_A \notin \mathcal{I}$. $\qquad\square$

Using Proposition 1.39, we get the following two corollaries.

Corollary 1.41 *Any proper ideal of a ring is contained in a maximal ideal.*

Corollary 1.42 *An element of a ring is invertible if and only if it is not contained in any maximal ideal of the ring.*

Definition 1.43

(i) *A ring with only one maximal ideal is local.*

(ii) *If A and B are local rings with respective maximal ideals $\mathcal{M}_A$ and $\mathcal{M}_B$, a homomorphism $f : A \to B$ such that $f(\mathcal{M}_A) \subset \mathcal{M}_B$ is called a local homomorphism of local rings.*

Exercises 1.44

1. Show that the ring $\mathbb{Z}_{(p)}$ (defined in Exercises 1.2) is local and that its maximal ideal is the set of all n/m with $n \in p\mathbb{Z}$ and $m \notin p\mathbb{Z}$.

2. Let K be a field and $(x_1, ..., x_n) \in K^n$. Show that the ring formed by all rational functions P/Q, with $P, Q \in K[X_1, ..., X_n]$ and such that $Q(x_1, ..., x_n) \neq 0$ is a local ring and that its maximal ideal is the set of all P/Q with $P(x_1, ..., x_n) = 0$ and $Q(x_1, ..., x_n) \neq 0$.

Definition 1.45 *The spectrum $\operatorname{Spec}(A)$ of a ring A is the set of all prime ideals of A.*

Proposition 1.46 *(Zariski topology)*
If, for each ideal $\mathcal{I}$ of A, we denote by $V(\mathcal{I}) \subset \operatorname{Spec}(A)$ the set of all prime ideals $\mathcal{P}$ such that $\mathcal{I} \subset \mathcal{P}$, the subsets $V(\mathcal{I})$ of $\operatorname{Spec}(A)$ are the closed sets of a topology on $\operatorname{Spec}(A)$.

Proof If $\mathcal{I}_s$, with $s = 1, ..., n$, are ideals of A, then $\bigcup_1^n V(\mathcal{I}_s) = V(\bigcap_1^n \mathcal{I}_s)$.
If $\mathcal{I}_s$ is a family of ideals of A, then $\bigcap_s V(\mathcal{I}_s) = V(\sum_s \mathcal{I}_s)$.   □

Note that by Corollary 1.41, a non-empty closed set of $\operatorname{Spec}(A)$ contains a maximal ideal.

Definition 1.47 *If $s \in A$, we denote by $D(s)$ the open set $\operatorname{Spec}(A) \setminus V(sA)$ of $\operatorname{Spec}(A)$.*

We recall that a closed set of a topological space is irreducible if it is not the union of two strictly smaller closed subsets.

Proposition 1.48 *If $\mathcal{P}$ is a prime ideal of a ring A, the closed set $V(\mathcal{P}) \subset \operatorname{Spec}(A)$ is irreducible.*

Proof Let F_1 and F_2 be closed sets of $\operatorname{Spec}(A)$ such that $V(\mathcal{P}) = F_1 \cup F_2$. Then there exists i such that $\mathcal{P} \in F_i$. Consequently, we have $V(\mathcal{P}) = F_i$.   □

As a special case, we get the following result.

Proposition 1.49 *Let A be a domain.*

(i) *The topological space* Spec(A) *is irreducible.*

(ii) *Any non-empty open subset of* Spec(A) *is dense in* Spec(A).

As we saw, the proof of this statement is straightforward. Its main consequence, which we will understand in due time, is that algebraic varieties are irreducible topological spaces for the Zariski topology.

1.4   Nilradicals and Jacobson radicals

Proposition 1.50 *The nilpotent elements of a ring A form an ideal of A.*

Proof   If $a^n = 0$ and $b^m = 0$, then $(ca - db)^{n+m-1} = 0$.   □

Corollary 1.51 *If $\mathcal{I}$ is an ideal of A, the set of all elements $a \in A$ having a power in $\mathcal{I}$ is an ideal of A.*

Proof   Apply the proposition to the ring $A/\mathcal{I}$.   □

Definition 1.52 *This ideal is the radical $\sqrt{\mathcal{I}}$ of $\mathcal{I}$. The radical $\sqrt{(0)}$ of (0) is the nilradical* Nil(A) *of A.*

Note that $\sqrt{\sqrt{\mathcal{I}}} = \sqrt{\mathcal{I}}$. As a consequence we see that if A is not the zero ring, then $A/\sqrt{(0)}$ is reduced.

Proposition 1.53 *A non-zero ring A is reduced if and only if* Nil(A) $= (0)$.

This is the definition of a reduced ring.

Theorem 1.54 *If A is not the zero ring, the nilradical* Nil(A) *is the intersection of all prime ideals of A.*

Proof   Consider $a \in$ Nil(A) and $\mathcal{P}$ a prime ideal. There exists $n > 0$ such that $a^n = 0$, hence $a^n \in \mathcal{P}$ and $a \in \mathcal{P}$. This proves Nil(A) $\subset \mathcal{P}$.

Assume now $a \notin$ Nil(A) and let us show that there exists a prime ideal $\mathcal{P}$ such that $a \notin \mathcal{P}$. Consider the part S of A consisting of all positive powers of a. We have assumed that $0 \notin S$. We can therefore consider the non-empty set E of all ideals of A which do not intersect S. Let E' be a totally ordered subset of E. Clearly E' is bounded above, in E, by $\bigcup_{\mathcal{I} \in E'} \mathcal{I}$. By Zorn's lemma, E has a maximal element.

If $\mathcal{I}$ is a maximal element in E, let us show that $\mathcal{I}$ is a prime ideal.

Let $x, y \in A$ be such that $xy \in \mathcal{I}$. If $x \notin \mathcal{I}$ and $y \notin \mathcal{I}$, there are positive integers n and m such that $a^n \in \mathcal{I} + xA$ and $a^m \in \mathcal{I} + yA$. This implies

$$a^{n+m} \in \mathcal{I} + x\mathcal{I} + y\mathcal{I} + xyA \subset \mathcal{I},$$

hence a contradiction.   □

Corollary 1.55 *If $\mathcal{I}$ is a proper ideal of A, the intersection of all prime ideals containing $\mathcal{I}$ is $\sqrt{\mathcal{I}}$.*

Proof Apply the theorem to the ring $A/\mathcal{I}$. □

As an important but straightforward consequence, we get the following result.

Corollary 1.56 *Let $\mathcal{I}$ and $\mathcal{J}$ be ideals of a ring A. The closed sets $V(\mathcal{I})$ and $V(\mathcal{J})$ of $\mathrm{Spec}(A)$ are equal if and only if $\sqrt{\mathcal{I}} = \sqrt{\mathcal{J}}$.*

Definition 1.57 *The intersection of all maximal ideals of a non-zero ring A is the Jacobson radical $\mathrm{JR}(A)$ of A.*

Theorem 1.58 *An element $a \in A$ is contained in $\mathrm{JR}(A)$ if and only if $1 - ax$ is invertible for all $x \in A$.*

Proof Assume $a \in \mathrm{JR}(A)$. If $\mathcal{M}$ is a maximal ideal, then $ax \in \mathcal{M}$, hence $1 - ax \notin \mathcal{M}$ (since $1 = (1 - ax) + ax$). Since $1 - ax$ is not contained in any maximal ideal, this is an invertible element.

Assume now $a \notin \mathrm{JR}(A)$ and let $\mathcal{M}$ be a maximal ideal such that $a \notin \mathcal{M}$. Since $A/\mathcal{M}$ is a field $\mathrm{cl}(a) \in A/\mathcal{M}$ is invertible. Hence there exists $b \in A$ such that $\mathrm{cl}(a)\mathrm{cl}(b) = \mathrm{cl}(1)$. In other words $\mathrm{cl}(1 - ab) = 0$. But this implies $(1 - ab) \in \mathcal{M}$ and $1 - ab$ is not invertible. □

1.5 Comaximal ideals

Definition 1.59 *We say that two ideals $\mathcal{I}$ and $\mathcal{J}$ of A are comaximal if*

$$\mathcal{I} + \mathcal{J} = A.$$

Note that $\mathcal{I}$ and $\mathcal{J}$ are comaximal if and only if the closed sets $V(\mathcal{I})$ and $V(\mathcal{J})$ of $\mathrm{Spec}(A)$ are disjoint. Indeed, $\mathcal{I} + \mathcal{J} = A$ if and only if there is no prime ideal $\mathcal{P}$ in $V(\mathcal{I}) \cap V(\mathcal{J})$.

Lemma 1.60 *Let $\mathcal{I}_l$, with $l = 1, ..., n$, be ideals pairwise comaximal. Then $\mathcal{I}_e$ and $\bigcap_{i \neq e} \mathcal{I}_i$ are comaximal for any $e \in [1, ..., n]$.*

Proof Assume $\mathcal{M}$ is a maximal ideal containing $\mathcal{I}_e$ and $\bigcap_{i \neq e} \mathcal{I}_i$. Since

$$\bigcap_{i \neq e} \mathcal{I}_i \subset \mathcal{M},$$

we know, from Proposition 1.36, that there exists $l \neq e$ such that $\mathcal{I}_l \subset \mathcal{M}$. But $\mathcal{I}_e$ and $\mathcal{I}_l$ are comaximal. This is a contradiction. □

Lemma 1.61 *If $\mathcal{I}_l$, with $l = 1, ..., n$, are pairwise comaximal, then*

$$\bigcap_1^n \mathcal{I}_l = \mathcal{I}_1 \mathcal{I}_2 ... \mathcal{I}_n.$$

Proof By the preceding lemma, we can assume $n = 2$. Let $1 = a + b$, with $a \in \mathcal{I}_1$ and $b \in \mathcal{I}_2$. If $x \in \mathcal{I}_1 \cap \mathcal{I}_2$, then $x = ax + bx \in \mathcal{I}_1 \mathcal{I}_2$. Hence $\mathcal{I}_1 \mathcal{I}_2 \subset \mathcal{I}_1 \cap \mathcal{I}_2 \subset \mathcal{I}_1 \mathcal{I}_2$. $\qquad\square$

Theorem 1.62 *Let $\mathcal{I}_1, ..., \mathcal{I}_n$ be ideals of A.*
The ideals $\mathcal{I}_l$ are pairwise comaximal if and only if the natural injective homomorphism

$$f : A/(\bigcap_l \mathcal{I}_l) \to \prod_1^n A/\mathcal{I}_l$$

is an isomorphism.

Proof Assume first that f is surjective. There exists $a \in A$ such that

$$\mathrm{cl}_1(a) = \mathrm{cl}_1(1) \in A/\mathcal{I}_1 \quad \text{and} \quad \mathrm{cl}_2(a) = \mathrm{cl}_2(0) \in A/\mathcal{I}_2.$$

This shows $1 - a \in \mathcal{I}_1$ and $a \in \mathcal{I}_2$, hence $1 = (1 - a) + a \in \mathcal{I}_1 + \mathcal{I}_2$. We have proved that for any $l \neq k$, the ideals $\mathcal{I}_l$ and $\mathcal{I}_k$ are comaximal.

Assume now that the ideals $\mathcal{I}_l$ are pairwise comaximal. By Lemma 1.60, one can find $a_e \in \mathcal{I}_e$ and $b_e \in \bigcap_{k \neq e} \mathcal{I}_k$ such that $1 = a_e + b_e$, for all e. In other words,

$$\mathrm{cl}_e(b_e) = \mathrm{cl}_e(1) \in A/\mathcal{I}_e \quad \text{and} \quad \mathrm{cl}_k(b_e) = 0 \in A/\mathcal{I}_k \quad \text{for} \quad k \neq e.$$

This shows

$$\mathrm{cl}_l(b_1 z_1 + b_2 z_2 + \cdots + b_n z_n) = \mathrm{cl}_l(z_l) \in A/\mathcal{I}_l$$

and the surjectivity of f. $\qquad\square$

Exercise 1.63 If $n = p_1^{r_1} ... p_m^{r_m}$ is a prime decomposition of the integer n, show that there is a natural isomorphism $\mathbb{Z}/n\mathbb{Z} \simeq \mathbb{Z}/p_1^{r_1}\mathbb{Z} \times \cdots \times \mathbb{Z}/p_m^{r_m}\mathbb{Z}$.

1.6   Unique factorization domains (UFDs)

Definition 1.64 *A non-invertible element of a ring is called irreducible if it is not the product of two non-invertible elements.*

Definition 1.65 *A domain A is a unique factorization domain if*

(i) *for all irreducible elements a of A the ideal aA is prime,*

(ii) *any non-zero and non-invertible element of A is a product of irreducible elements.*

Theorem 1.66 *(The uniqueness of the decomposition as a product of irreducible elements)*
   Let A be a UFD. If $a_i(i = 1, ..., n)$ and $b_j(j = 1, ..., m)$ are irreducible elements of A such that $\prod_1^n a_i = \prod_1^m b_j$, then:

(i) $n = m;$

(ii) *there is a permutation τ of $[1, ..., n]$ such that $a_i A = b_{\tau(i)} A$ for all i.*

Proof Since $a_i A$ is a prime ideal, there exists j such that $b_j \in a_i A$. Let $b_j = a_i c$. Since b_j is irreducible, c is invertible. Hence $b_j A = a_i A$. The theorem follows. □

Definition 1.67 *Let A be a UFD and $a, b \in A$ non-zero elements.*

(i) *A divisor $d \in A$ of a and b such that any common divisor of a and b is a divisor of d is called a gcd (greatest common divisor) of a and b.*
   If 1 is a gcd of a and b, we say that a and b are relatively prime.

(ii) *A multiple $m \in A$ of a and b such that any common multiple of a and b is a multiple of m is called a lcm (least common multiple) of a and b.*

The proof of the following result is easy and left to the reader.

Proposition 1.68 *Let A be a UFD and $a, b \in A$ non-zero elements.*

(i) *The elements a and b have a gcd and a lcm.*

(ii) *If c is a gcd (resp. lcm) of a and b, then c' is a gcd (resp. lcm) of a and b if and only if $cA = c'A$.*

(iii) *If c (resp. m) is a gcd (resp. lcm) of a and b, then $abA = cmA$.*

CAREFUL: If A is a UFD and a and b are relatively prime elements of A, one does not necessarily have $aA + bA = A$.

Theorem 1.69 *A principal ideal ring is a UFD.*

   Our proof depends on the following lemma:

Lemma 1.70 *An increasing sequence of ideals in a principal ideal domain is stationary.*

Proof Let A be the ring and $a_1 A \subset a_2 A \subset \ldots \subset a_n A \subset \ldots$ such an increasing sequence. Then $\mathcal{I} = \bigcup_{i>0} a_i A$ is obviously an ideal of A. Consequently, there exists $b \in \mathcal{I}$ such that $\mathcal{I} = bA$. But $b \in a_i A$ for i large enough. This shows $\mathcal{I} = a_i A$ for i large enough, hence the lemma.   □

Proof of Theorem 1.69

Let a be an irreducible element of a principal ideal domain A. Let $\mathcal{M}$ be a maximal ideal containing a and $b \in \mathcal{M}$ a generator of $\mathcal{M}$. If $a = bc$, then c is invertible, since a is irreducible. This proves $aA = bA = \mathcal{M}$, hence aA is prime.

Assume $a_1 \in A$ is not a product of irreducible elements. Let $\mathcal{M} = bA$ be a maximal ideal containing a_1. There exists a_2 such that $a = ba_2$. We found $a_2 \in A$ such that $a_1 A \subset a_2 A$, $a_1 A \neq a_2 A$ and a_2 is not a product of irreducible elements. As a consequence, if there exists a non-zero element which is not a product of irreducible elements, we can construct an infinite strictly increasing sequence of ideals in A. This contradicts our lemma.   □

Exercise 1.71 Let a and b be non-zero elements of a principal ideal ring A. Prove that they are relatively prime if and only if $aA + bA = A$.

CAREFUL : As we noted before, this result is not true in a UFD which is not a principal ideal ring.

To end this chapter, we state a fundamental result which we will prove in chapter 7. It is certainly possible to prove it already here and now, but it will appear in Corollary 7.57 as a special case of a theorem concerning fractions and UFDs.

Theorem 1.72 *If A is a UFD (or a field), the polynomial ring $A[X]$ is a UFD.*

Corollary 1.73 *A polynomial ring $K[X_1, \ldots, X_n]$ over a field K is a UFD.*

1.7   Exercises

1. Show that if a ring A has only one prime ideal, then an element of A is invertible or nilpotent.

2. Let K_i, with $i = 1, \ldots, n$, be fields. Show that the ring $K_1 \times \cdots \times K_n$ has only finitely many ideals.

3. If A is a principal ideal ring and $a \in A$ a non-zero element, show that the quotient ring A/aA has only finitely many ideals.

4. Let A be a UFD and $a \in A$ a non-zero element. Show that the nilradical of A/aA is the intersection of a finite number of prime ideals. If $\mathcal{P}_1, \ldots, \mathcal{P}_n$ are these prime ideals, show that for each prime ideal $\mathcal{P}$ of A/aA there exists i such that $\mathcal{P}_i \subset \mathcal{P}$.

5. Let A be a ring and $a \in A$ a nilpotent element. Show that $1 + a$ is invertible. If $a^n = 0$ and $a^{n-1} \neq 0$ describe the inverse of a.

6. In a ring A, let e and e' be non-zero elements such that $1 = e + e'$ and $ee' = 0$. Show that A is the product of two rings.

7. Let $\mathcal{P}$ be a prime ideal of a ring A. Show that the ideal $\mathcal{P}A[X]$ of the polynomial ring $A[X]$ is prime.

8. Let $\mathrm{Nil}(A)$ be the nilradical of a ring A. Show that $\mathrm{Nil}(A)A[X]$ is the nilradical of the polynomial ring $A[X]$.

2

Modules

As in the preceding chapter, we start with some fairly unexciting results. In the first three sections we introduce the basic language and notations; we will be moving fast. But reader, please, be sure to understand the statements and do the exercises. In section 4, we meet free modules and many pleasant memories from linear algebra over a field. Simultaneously we will make a sad discovery: so many modules are not free. In other words, when studying finitely generated modules, bases are not always available. In section 6 we see that this lack doesn't hinder us from using matrices. They are essential in proving Nakayama's lemma and a generalization of the Cayley–Hamilton theorem, two results often used later on.

Throughout this chapter A is a given ring.

Definition 2.1 *An A-module M is a commutative group equipped with a map*

$$A \times M \to M; \quad (a, x) \to ax,$$

satisfying the following relations for all $x, y \in M$ and $a, b \in A$:

(i) $a(x + y) = ax + ay,$

(ii) $(a + b)x = ax + bx,$

(iii) $a(bx) = (ab)x,$

(iv) $1x = x.$

Examples 2.2

1. Obviously A is an A-module.

2. An ideal of A is an A-module.

3. If $\mathcal{I}$ is an ideal of A, the quotient $A/\mathcal{I}$ has a natural structure of an A-module, defined by $a\mathrm{cl}(x) = \mathrm{cl}(ax)$.

4. If $f : A \to B$ is a ring homomorphism, B has a natural structure of an A-module, defined by $ab = f(a)b$. We often denote this A-module by $f_*(B)$.

17

2.1   Submodules. Homomorphisms. Quotient modules

Definition 2.3 *A subset N of M is an A-submodule of M if*

$$x, y \in N \quad \text{and} \quad a, b \in A \Rightarrow ax + by \in N.$$

When there is no ambiguity about the base ring, we will write submodule of M instead of A-submodule of M.

Examples 2.4

1. An ideal of A is a submodule of A.
2. If $x \in M$, then Ax is a submodule of M.
3. The A-submodules of $A/\mathcal{I}$ are the ideals of this quotient ring of A.

Definition 2.5 *If $x_1, ..., x_n \in M$, a linear combination of $x_1, ..., x_n$ is an element of the form $a_1x_1 + \cdots + a_nx_n$, with $a_1, ..., a_n \in A$.*

The next proposition is clear.

Proposition 2.6 *Let x_i $(i \in E)$ be elements of M. The set of all linear combinations of these elements is a submodule of M. It is the smallest submodule of M containing the elements x_i $(i \in E)$.*

We say that this submodule is generated by the elements x_i $(i \in E)$. If this submodule is M, then x_i $(i \in E)$ is a system of generators of M.

Definition 2.7 *Let M and N be A-modules. A map $f : M \to N$ is a homomorphism of A-modules (or is A-linear) if*

$$f(ax + by) = af(x) + bf(y) \quad \text{for all } x, y \in M \text{ and } a, b \in A.$$

Note that the composition of two composable homomorphisms is a homomorphism. Note furthermore that if a homomorphism is bijective, then the inverse map is a homomorphism (it is easy to check); such a homomorphism is an isomorphism.

Proposition 2.8 *Let $f : M \to N$ be a homomorphism of A-modules. The kernel $f^{-1}(0)$ of f is an A-submodule of M, denoted $\ker f$. The image $f(M)$ of f is an A-submodule of N.*

Check it!

Theorem 2.9 *Let N be an submodule of M. The quotient commutative group M/N (the group of equivalence classes for the relation $x \sim y$ if $(x-y) \in N$) has a unique structure of an A-module such that the class map $\mathrm{cl} : M \to M/N$ is A-linear (in other words $a\,\mathrm{cl}(x) + b\,\mathrm{cl}(y) = \mathrm{cl}(ax+by)$).*
The kernel of this map is N.

Proof If $(x-x') \in N$ and $(y-y') \in N$, then $((ax+by)-(ax'+by')) \in N$ for all $a, b \in A$. It is therefore possible to define

$$a\,\mathrm{cl}(x) + b\,\mathrm{cl}(y) = \mathrm{cl}(ax+by)$$

and we are done. □

As for ring homomorphisms, we have a natural factorization which we will use often later on. The proof is easy but the result important.

Theorem 2.10 *(The factorization theorem)*
If $f : M \to N$ is a homomorphism of A-modules, there exists a unique isomorphism $\overline{f} : M/\ker f \simeq f(M)$ such that $f = \mathrm{i} \circ \overline{f} \circ \mathrm{cl}$ where $\mathrm{cl} : M \to M/\ker f$ and $\mathrm{i} : f(M) \to N$ are the natural applications.

Proof Since $(x-y) \in \ker f$ implies $f(x) = f(y)$, we can define $\overline{f}(\mathrm{cl}(x)) = f(x)$ and check immediately the factorization. □

Definition 2.11 $N/f(M)$ *is the cokernel,* $\mathrm{coker}\, f$, *of the homomorphism f.*

Proposition 2.12 *Let N be a submodule of M.*

(i) *If F is a submodule of M/N, then $\mathrm{cl}^{-1}(F)$ is a submodule of M containing N.*

(ii) *If N' is a submodule of M containing N, then $\mathrm{cl}(N') \simeq N'/N$ is a submodule of M/N.*

(iii) *We have $\mathrm{cl}^{-1}(\mathrm{cl}(N')) = N'$ and $\mathrm{cl}(\mathrm{cl}^{-1}(F)) = F$.*

(iv) *This bijection between the submodules of M containing N, and the submodules of M/N respects inclusion.*

The proof is straightforward. The following proposition helps.

Proposition 2.13 (i) *If F is a submodule of M/N, then $\mathrm{cl}^{-1}(F)$ is the kernel of the composition homomorphism*

$$M \to M/N \to (M/N)/F.$$

(ii) *This homomorphism factorizes through the isomorphism*

$$M/\mathrm{cl}^{-1}(F) \simeq (M/N)/F.$$

This description of $\mathrm{cl}^{-1}(F)$ needs no comment. The factorization is a consequence of the factorization theorem.

Definition 2.14 *Let N be a submodule of M and $P \subset M$ a subset.*
We denote by $N : P$ the set of all $a \in A$ such that $ax \in N$ for all $x \in P$. This is the conductor of P in N. In particular the annihilator of M is $\mathrm{ann}(M) = (0) : M$ (the conductor of M in (0)).

Note that $N : P$ is an ideal.

Exercise 2.15 For $x \in M$, show $Ax \simeq A/((0) : x)$.

Definition 2.16 *An A-module M is faithful if $(0) : M = (0)$.*

Exercise 2.17 Consider the $\mathbb{Z}$-module $\mathbb{Q}/\mathbb{Z}$. Show that $((0) : x) \neq (0)$ for all $x \in \mathbb{Q}/\mathbb{Z}$ but that $\mathbb{Q}/\mathbb{Z}$ is a faithful $\mathbb{Z}$-module.

2.2   Products and direct sums

Proposition 2.18 *Let $(M_i)_{i \in E}$ be a family of A-modules.*
The set-theoretical product $\prod_{i \in E} M_i = \{(x_i)_{i \in E}\}$, equipped with the following operations, is an A-module.

$$(x_i)_{i \in E} + (y_i)_{i \in E} = (x_i + y_i)_{i \in E} \quad \text{and} \quad a((x_i)_{i \in E}) = (ax_i)_{i \in E}.$$

Definition 2.19 *The direct sum $\bigoplus_{i \in E} M_i$ is the submodule of $\prod_{i \in E} M_i$ formed by all*

$$(x_i)_{i \in E}, \quad x_i = 0 \quad \text{for all } i \text{ except finitely many.}$$

Definition 2.20 (i) *We denote by nM, or $\bigoplus_1^n M$, or M^n the direct sum of n copies of M.*

(ii) *More generally, if E is a set and if $M_i = M$ for all $i \in E$, we denote by $\bigoplus_{i \in E} M$ the direct sum of the A-modules M_i.*

One has, obviously:

Proposition 2.21 *If E is finite, then $\bigoplus_{i \in E} M_i = \prod_{i \in E} M_i$.*

Exercise 2.22 Let $(M_i)_{i \in E}$ be a family of A-modules and, for each $i \in E$, let N_i be a submodule of M_i. Note first that $\bigoplus_{i \in E} N_i$ is naturally a submodule of $\bigoplus_{i \in E} M_i$ and then show that there is a natural isomorphism

$$\left(\bigoplus_{i \in E} M_i \right) / \left(\bigoplus_{i \in E} N_i \right) \simeq \bigoplus_{i \in E} (M_i/N_i).$$

2.3 Operations on the submodules of a module

Definition 2.23 *Let N and N' be submodules of M. Their sum is*

$$N + N' = \{x + x', \ x \in N, \ x' \in N'\}.$$

Obviously, $N + N'$ is the smallest submodule of M containing N and N'.

Definition 2.24 *If N_i ($i \in E$) are submodules of M, we denote by $\sum_s N_i$ the smallest submodule of M containing N_i for all $i \in E$.*

It is clear that $x \in \sum_s N_i$ if and only if there exist $x_i \in N_i$, all zeros except for a finite number, such that $x = \sum_i x_i$.

DANGER: $N \cup N'$ is not generally a submodule.

Definition 2.25 *Let $\mathcal{I}$ be an ideal of A. We denote by $\mathcal{I}M$ the submodule of M formed by all linear combinations of elements of M with coefficients in $\mathcal{I}$.*

The proof of the following result is straightforward.

Proposition 2.26 *Let N and N' be submodules of M.*

(i) *There is a natural surjective homomorphism $N \oplus N' \to N + N'$ whose kernel is isomorphic to $N \cap N'$.*

(ii) *If $N \cap N' = (0)$, there is a natural isomorphism $N \oplus N' \simeq N + N'$.*

(iii) *If furthermore $N + N' = M$, then $M \simeq N \oplus N'$. In this case we say that N and N' are direct factors of M.*

Note that there are in fact several natural surjective homomorphisms $N \oplus N' \to N + N'$ whose kernel is isomorphic to $N \cap N'$:

$$x \oplus x' \to x + x' \quad \text{and} \quad x \oplus x' \to x - x'$$

are typical examples.

We conclude this section with an easy but particularly convenient result.

Theorem 2.27 *(The isomorphism theorem)*

(i) *Let $N \subset N' \subset M$ be submodules of M. There is a natural isomorphism*

$$M/N' \simeq (M/N)/(N'/N).$$

(ii) *Let N and F be submodules of M. There is a natural isomorphism*

$$N/(N \cap F) \simeq (N + F)/F.$$

Proof For (i), check that the kernel of the surjective composition homomorphism

$$M \to M/N \to (M/N)/(N'/N)$$

is N', and use the factorization theorem.

For (ii), show that $N \cap F$ is the kernel of the natural surjective homomorphism

$$N \to (N + F)/F,$$

and use once more the factorization theorem.   □

2.4   Free modules

Definition 2.28 *Let x_i, with $i \in E$, be elements of M.*

(i) *If the homomorphism*

$$\bigoplus_{i \in E} A \to M, \quad (a_i)_{i \in E} \to \sum_{i \in E} a_i x_i$$

is injective, we say that the elements x_i, $i \in E$, are linearly independent.

(ii) *If the elements x_i, $i \in E$, are linearly independent and generate M, they form a basis of M.*

(iii) *An A-module with a basis is a free module.*

Note that (x_i), with $i \in E$, is a basis of M if and only if every element in M has a unique decomposition as a linear combination of the elements x_i, $i \in E$.

Example 2.29 The direct sum nA is a free A-module.

Proposition 2.30 *If a free A-module L has a finite basis, all its bases have the same number of elements. This number is the rank of the free module L.*

Proof When A is a field, this statement is well known. Since every ring has a maximal ideal, the proposition is a consequence of the following lemma applied to a maximal ideal.

Lemma 2.31 *Let L be a free A-module and $(e_1, ..., e_n)$ a basis of L. If $\mathcal{I}$ is an ideal of A, then $L/\mathcal{I}L$ is a free $A/\mathcal{I}$-module and the elements*

$$\mathrm{cl}(e_1), ..., \mathrm{cl}(e_n) \in L/\mathcal{I}L$$

form a basis of $L/\mathcal{I}L$.

Proof It is clear that $(\mathrm{cl}(e_1), ..., \mathrm{cl}(e_n))$ is a system of generators of $L/\mathcal{I}L$. Consider a relation $\sum_{i=1}^n \mathrm{cl}(a_i)\mathrm{cl}(e_i) = 0$, with coefficients $\mathrm{cl}(a_i) \in A/\mathcal{I}$. This relation can be written $\sum_{i=1}^n a_i e_i \in \mathcal{I}L$. In other words there exist $b_i \in \mathcal{I}$ such that $\sum_{i=1}^n a_i e_i = \sum_{i=1}^n b_i e_i$. Since $(e_1, ..., e_n)$ is a basis of L, it implies $a_i = b_i$, hence $\mathrm{cl}(a_i) = 0$, for all i. □

Be careful with habits from linear algebra over a field. If you have n linearly independent elements in a free module M of rank n, they do not necessarily form a basis of M (find an example). But we see with the next statement, that n elements generating M do.

Proposition 2.32 *Let L be a free A-module of rank n. If $x_1, ..., x_n \in L$ generate L, then $(x_1, ..., x_n)$ is a basis.*

Proof Let $(e_1, ..., e_n)$ be a basis of L. There are $n \times n$ matrices S and T with coefficients in A, such that

$$T \begin{pmatrix} x_1 \\ x_2 \\ . \\ . \\ x_n \end{pmatrix} = \begin{pmatrix} e_1 \\ e_2 \\ . \\ . \\ e_n \end{pmatrix} \quad \text{and} \quad S \begin{pmatrix} e_1 \\ e_2 \\ . \\ . \\ e_n \end{pmatrix} = \begin{pmatrix} x_1 \\ x_2 \\ . \\ . \\ x_n \end{pmatrix}.$$

The relation $TS \begin{pmatrix} e_1 \\ e_2 \\ . \\ . \\ e_n \end{pmatrix} = \begin{pmatrix} e_1 \\ e_2 \\ . \\ . \\ e_n \end{pmatrix}$ implies $TS = I_{n \times n}$. Hence the determinant of S is invertible and S is an invertible matrix.

Consider a relation $a_1 x_1 + \cdots + a_n x_n = 0$. It induces a relation

$$(a_1, ..., a_n)S \begin{pmatrix} e_1 \\ e_2 \\ . \\ . \\ e_n \end{pmatrix} = 0,$$

hence a relation $(a_1, ..., a_n)S = 0$. Since S is invertible, this proves $a_i = 0$ for all i. □

CAREFUL: If A is not a field, there are many non-free A-modules. For example $A/\mathcal{I}$ where $\mathcal{I}$ is a non-zero ideal.

Exercises 2.33

1. Show that a non-zero ideal of A is free if and only if it is principal and generated by a non-zero divisor of A.

2. Let L be a free A-module of rank n and $(e_1, ..., e_n)$ a basis for L. Consider a positive integer $m \leq n$ and elements $a_1, ..., a_m \in A$. If M is the submodule of L generated by $a_1 e_1, ..., a_m e_m$, show that

$$L/M \simeq \bigoplus_1^m A/a_i A \bigoplus (n-m)A.$$

2.5   Homomorphism modules

Definition 2.34 *Let M and N be A-modules. The set $\mathrm{Hom}_A(M, N)$ of all homomorphisms from M to N is an A-module.*

(i) *If $f \in \mathrm{Hom}_A(M, N)$ and $a \in A$, then $(af)(x) = af(x) = f(ax)$.*

(ii) *If $h : N \to N'$ is a homomorphism, $\mathrm{Hom}_A(M, h) : \mathrm{Hom}_A(M, N) \to \mathrm{Hom}_A(M, N')$ is the homomorphism defined by $\mathrm{Hom}_A(M, h)(f) = h \circ f$.*

(iii) *If $g : M' \to M$ is a homomorphism, $\mathrm{Hom}_A(g, N) : \mathrm{Hom}_A(M, N) \to \mathrm{Hom}_A(M', N)$ is the homomorphism defined by $\mathrm{Hom}_A(g, N)(f) = f \circ g$.*

DANGER: There exist non-zero modules M and N such that

$$\mathrm{Hom}_A(M, N) = (0).$$

For example, if $a \in A$ is a non-zero divisor, then $\mathrm{Hom}_A(A/aA, A) = (0)$. Indeed, let $f \in \mathrm{Hom}_A(A/aA, A)$ and $y \in A/aA$. We have $af(y) = f(ay) = 0$, hence $f(y) = 0$.

Definition 2.35 *The module $M\widetilde{\,} = \mathrm{Hom}_A(M, A)$ is the dual of M.*

Note that we have described a non-zero A-module whose dual is zero.

Proposition 2.36 *Let M be an A-module. The evaluation homomorphism $e : M \to M\widetilde{\widetilde{\,}}$, defined by $e(x)(f) = f(x)$, is linear.*
If this homomorphism is an isomorphism, we say that M is reflexive.

Proposition 2.37 *Let L be a free A-module of rank n, then:*

(i) *$L\widetilde{\,}$ is free of rank n;*

(ii) *L is reflexive.*

Proof Let $(f_i)_{i\in I}$ be a basis of L. We define, as for vector spaces over a field, the dual basis $(\tilde{f_i})_{i\in I}$ of $L\check{\,}$:

$$\tilde{f_i}(f_j) = \delta_{ij}.$$

Verifying that it is a basis of $L\check{\,}$ is done exactly as for vector spaces. Then $(\tilde{f_i}^{\check{}})_{i\in I}$ is a basis of $L\check{\,}\check{\,}$. Checking $e(f_i) = \tilde{f_i}^{\check{}}$ is straightforward. □

Exercises 2.38 A reflexive module is not necessarily free.

1. Show that all ideals of $\mathbb{Z}/n\mathbb{Z}$ are reflexive (but not all free) $\mathbb{Z}/n\mathbb{Z}$-modules.

2. Show that $(X_0, X_1)/(X_0X_3 - X_1X_2)$ is a reflexive but not free ideal of the ring

$$\mathbb{C}[X_0, X_1, X_2, X_3]/(X_0X_3 - X_1X_2).$$

2.6 Finitely generated modules

In this last section, we focus for the first time on finitely generated modules. We will often come back to such modules later on. Two extremely useful results are established here. The first, Nakayama's lemma, is a convenient criterion in deciding whether a finitely generated module is zero or not. The second, a generalization of the Cayley–Hamilton theorem, will be particularly suitable, from chapter 8 on, for finding algebraic relations. The proofs make systematic use of matrices with coefficients in a ring and are elementary.

Definition 2.39 *An A-module generated by a finite number of elements is of finite type (or finitely generated).*

Proposition 2.40 *Let M be an A-module and N a submodule of M.*

(i) *If M is finitely generated, so is M/N.*

(ii) *If N and M/N are finitely generated, so is M.*

Proof (i) If $x_1, ..., x_n \in M$ generate M, then it is clear that $\mathrm{cl}(x_1), ..., \mathrm{cl}(x_n) \in M/N$ generate M/N.

(ii) Assume now that $x_1, ..., x_n \in N$ generate N and that $\mathrm{cl}(y_1), ..., \mathrm{cl}(y_m) \in M/N$ generate M/N. We claim that the elements $x_1, ..., x_n, y_1, ..., y_m$ of M generate M. Indeed, if $z \in M$, then there is, in M/N, a decomposition $\mathrm{cl}(z) = \sum_i a_i\mathrm{cl}(y_i)$. This implies $z - (\sum_i a_iy_i) \in N$ and we are done. □

Theorem 2.41 *(Nakayama's lemma)*

Let $\mathcal{J}$ be contained in the Jacobson radical of A. If M is a finitely generated module such that $M = \mathcal{J}M$, then $M = (0)$.

Proof Assume $(x_1, ..., x_n)$ generate M. Since $M = \mathcal{J}M$, there exist $a_{ij} \in \mathcal{J}$ such that $x_i = \sum_j a_{ij}x_j$. Consider the $n \times n$ matrix, with coefficient in the ring, $T = I_{n\times n} - (a_{ij})$. We have

$$T \begin{pmatrix} x_1 \\ x_2 \\ . \\ . \\ x_n \end{pmatrix} = 0.$$

Let ${}^t\mathrm{Co}(T)$ be the transpose of the cofactor matrix of T. We get

$$^t\mathrm{Co}(T)T \begin{pmatrix} x_1 \\ x_2 \\ . \\ . \\ x_n \end{pmatrix} = 0,$$

hence

$$\det(T) \begin{pmatrix} x_1 \\ x_2 \\ . \\ . \\ x_n \end{pmatrix} = 0.$$

In other words $\det(T)x_i = 0$ for all i, hence $\det(T)M = 0$. But we have $\det(T) = 1 + a$, with $a \in \mathcal{J}$. Since, by Theorem 1.58, $1 + a$ is invertible, this implies $M = (0)$. □

Corollary 2.42 *Let M be a finitely generated A-module and $\mathcal{J}$ an ideal contained in the Jacobson radical of A. The elements $x_1, ..., x_n \in M$ generate M if and only if $\mathrm{cl}(x_1), ..., \mathrm{cl}(x_n) \in M/\mathcal{J}M$ generate $M/\mathcal{J}M$.*

Proof One implication is obvious. Assume that $\mathrm{cl}(x_1), ..., \mathrm{cl}(x_n) \in M/\mathcal{J}M$ generate $M/\mathcal{J}M$.

Consider $x \in M$. There exist $a_1, ..., a_n \in A$ such that $\mathrm{cl}(x) = \sum a_i \mathrm{cl}(x_i)$. This implies $(x - (\sum a_i x_i)) \in \mathcal{J}M$. Let N be the sub-module of M generated by $x_1, ..., x_n$. We have proved $M/N = \mathcal{J}(M/N)$, hence $M/N = (0)$ by Nakayama's lemma. □

Exercise 2.43 Let A be a ring and $a \in A$ an element contained in the Jacobson radical of A. For all A-modules F, we denote $\mathrm{cl}_F : F \to F/aF$ the natural application. Let $f : N \to M$ be a homomorphism of finitely generated A-modules and $\overline{f} : N/aN \to M/aM$ the homomorphism such that $\overline{f}(\mathrm{cl}_N(x)) = \mathrm{cl}_M(f(x))$.

1. Show that f is surjective if and only if $\overline{f}$ is surjective.

2. Assume that $x \in M$ and $ax = 0$ imply $x = 0$ and that $\overline{f}$ is injective. Show that for all $y \in \ker f$ there exists $y' \in \ker f$ such that $y = ay'$. Prove then that if $\ker f$ is a finitely generated A-module, f is injective.

Theorem 2.44 *(Cayley–Hamilton revisited) Let $(x_1, ..., x_n)$ be a system of generators of an A-module M.*

If u is an endomorphism of M, let (a_{ij}) be a $(n \times n)$-matrix, with coefficients in A such that $u(x_i) = \sum_j a_{ij} x_j$ for all i.

Then, if $P(X) = \det(X I_{n \times n} - (a_{ij})) \in A[X]$, the endomorphism $P(u)$ of M is trivial.

Proof Let us give to M the structure of an $A[X]$-module by defining $Xy = u(y)$ for all $y \in M$. We get

$$(X I_{n \times n} - (a_{ij})) \begin{pmatrix} x_1 \\ x_2 \\ . \\ . \\ x_n \end{pmatrix} = 0.$$

Multiplying on the left by ${}^t\mathrm{Co}(X I_{n \times n} - (a_{ij}))$, we find

$$\det(X I_{n \times n} - (a_{ij})) \begin{pmatrix} x_1 \\ x_2 \\ . \\ . \\ x_n \end{pmatrix} = 0.$$

This means $\det(X I_{n \times n} - (a_{ij})) x_i = 0$ for all i, hence

$$\det(X I_{n \times n} - (a_{ij})) M = (0).$$

In other words $P(u) = 0$, by definition of the operation of X in M. $\square$

Exercise 2.45 Let B be an A-algebra. Assume that B is a finitely generated A-module. If $x \in B$, show that there exists a monic polynomial $P \in A[X]$ such that $P(x) = 0$.

2.7 Exercises

1. Let M_1 and M_2 be submodules of an A-module M. Show that the natural injective homomorphism $M/(M_1 \cap M_2) \to M/M_1 \oplus M/M_2$ is an isomorphism if and only if $M = M_1 + M_2$.

2. Let $P = X^n + a_1 X^{n-1} + \cdots + a_n \in A[X]$ be a monic polynomial with coefficients in the ring A. Show that the A-algebra $B = A[X]/(P)$ is a free A-module of rank n and that $(\mathrm{cl}(X^0), \ldots, \mathrm{cl}(X^{n-1}))$ is a basis for this A-module.

   Consider the dual basis $(\mathrm{cl}(X^0)\check{\ }, \ldots, \mathrm{cl}(X^{n-1})\check{\ })$ of $B\check{\ } = \mathrm{Hom}_A(B, A)$. Note that $B\check{\ }$ is equipped with the structure of a B-module defined by $bf(x) = f(bx)$ for $b, x \in B$ and $f \in B\check{\ }$. Show that this B-module is free of rank one and that $\mathrm{cl}(X^{n-1})\check{\ }$ is a basis for it.

3. Let A be a local ring and $\mathcal{M}$ its maximal ideal. Show that a homomorphism $M \to N$ of finitely generated A-modules is surjective if and only if the induced composed homomorphism $M \to N/\mathcal{M}N$ is surjective.

4. Let A be a local ring and L and F be finite rank free A-modules. Consider bases $(e_1, \ldots, e_l)$ and $(f_1, \ldots, f_n)$ of L and F. To a homomorphism $u : L \to F$, associate the matrix $M = (a_{ij})$, with $u(e_j) = \sum_i a_{ij} f_i$.

   Show that u is surjective if and only if M has an invertible n-minor.

5. With the same hypothesis as in the preceding exercise, show that the ideal generated by the n-minors of M and the annihilator of coker u have the same radical.

6. With the same hypothesis as in the preceding exercise, show that coker u is free of rank $n - l$ if and only if M has an invertible l-minor.

7. Let u be an endomorphism of a finitely generated free A-module. If $N = \mathrm{coker}\ u$, show that there exists a principal ideal having the same radical as the annihilator $(0) : N$ of N (I know that I have already asked this question, but this is important and I want to be sure that you remember!).

8. Let L be a finitely generated free A-module and $(e_1, \ldots, e_n)$ a basis of L. Consider a relation $1 = \sum_{i=1}^n a_i b_i$ in A. If $f : L \to A$ is the homomorphism defined by $f(e_i) = a_i$, show that there exists $e \in L$ such that $L = Ae \oplus \ker f$.

3

Noetherian rings and modules

Complex algebraic geometry is basically the study of polynomials with coefficients in $\mathbb{C}$, with a special interest in their zeros. As a consequence, polynomial rings over $\mathbb{C}$, their quotient rings and their fraction rings (they will be defined in chapter 7) play a central role in this text. By Hilbert's Theorem, all these rings are Noetherian.

As a remarkable first consequence we find that an algebraic set in $\mathbb{C}^n$ (the common zeros to a family of polynomials $f_i \in \mathbb{C}[X_1, ..., X_n]$) is defined by a finite number of polynomials (every ideal of the polynomial ring is finitely generated).

The primary decomposition of an ideal in a Noetherian ring has also, although it is not as obvious, an important geometric consequence: an algebraic set is a finite union of irreducible (for the Zariski topology) algebraic sets (every ideal of the polynomial ring has a primary decomposition). This last statement will only be clear after understanding Hilbert's Nullstellensatz.

I hope that you agree with the following result. If not, prove it.

Proposition 3.1 *Let E be an ordered set. The following conditions are equivalent.*

(i) *Any non-empty subset of E has a maximal element.*

(ii) *Any increasing sequence of E is stationary.*

We should perhaps recall that a sequence (x_n) is stationary if there exists $n_0 \in \mathbb{Z}$ such that $x_n = x_{n_0}$ for $n \geq n_0$.

3.1   Noetherian rings

Definition 3.2 *If the set of ideals of a ring satisfies the equivalent conditions of the preceding proposition, the ring is Noetherian.*

Since the ideals of a quotient ring $A/\mathcal{I}$ are naturally in bijection with the ideals of A containing $\mathcal{I}$, we get the following proposition.

Proposition 3.3 *If $\mathcal{I}$ is an ideal of a Noetherian ring, the quotient ring $A/\mathcal{I}$ is Noetherian.*

Theorem 3.4 *A ring is Noetherian if and only if all its ideals are finitely generated.*

Proof Assume first that A is Noetherian. Let $\mathcal{I}$ be an ideal of A. Consider the set E of finitely generated ideals contained in $\mathcal{I}$. Let $\mathcal{J}$ be a maximal element of this set. If $a \in \mathcal{I}$, then the ideal $Aa + \mathcal{J}$ is in the set E. This shows $Aa + \mathcal{J} = \mathcal{J}$, hence $\mathcal{J} = \mathcal{I}$. Conversely, assume all ideals of A are finitely generated. Let $\mathcal{I}_l$, with $l \geq 0$, be an increasing sequence of ideals of A. Clearly $\mathcal{I} = \bigcup_{l \geq 0} \mathcal{I}_l$. is an ideal. Let $(a_1, ..., a_n)$ be a system of generators of $\mathcal{I}$. There is an integer r such that $a_1, ..., a_n \in \mathcal{I}_r$. This proves $\mathcal{I} = \mathcal{I}_r$ and obviously $\mathcal{I}_l = \mathcal{I}_r$ for $l > r$.   □

Examples 3.5

1. A field is a Noetherian ring.
2. A principal ideal ring is a Noetherian ring.

Note that we proved earlier (Lemma 1.70) that an increasing sequence of ideals in a principal ideal domain is stationary.

Our next result is elementary but fundamental. Think about it. Consider some polynomials $P_i \in \mathbb{C}[X_1, \ldots, X_n]$, perhaps infinitely many. Hilbert's theorem states that the ideal of $\mathbb{C}[X_1, \ldots, X_n]$ generated by the polynomials P_i is finitely generated, say by $Q_1, \ldots, Q_r$. Consequently if $E \subset \mathbb{C}^n$ is the set of all $(x_1, \ldots, x_n)$ such that $P_i(x_1, \ldots, x_n) = 0$ for all i, then

$$(x_1, \ldots, x_n) \in E \iff Q_1(x_1, \ldots, x_n) = \cdots = Q_r(x_1, \ldots, x_n) = 0.$$

In other words the set E is defined by a finite number of polynomials.

Theorem 3.6 *(Hilbert's theorem)*
If A is a Noetherian ring, the polynomial ring $A[X]$ is Noetherian.

Proof Let $\mathcal{J}$ be a non-zero ideal of $A[X]$. Consider, for all $n \geq 0$, the ideal $\mathcal{I}_n$, of A, whose elements are the leading coefficients of the polynomials of degree n contained in $\mathcal{J}$ (we recall that the polynomial 0 has all degrees). Note that since $P \in \mathcal{J}$ implies $XP \in \mathcal{J}$, the sequence $\mathcal{I}_n$ is increasing. Hence there exists m such that $\mathcal{I}_n = \mathcal{I}_m$ for $n \geq m$.

Consider, for all $i \leq m$, a system of generators $(a_{i1}, ..., a_{in_i})$ of $\mathcal{I}_i$. Let $P_{i1}, ..., P_{in_i} \in \mathcal{J}$ be polynomials such that $\deg(P_{ij}) = i$ and the leading coefficient of P_{ij} is a_{ij}. We claim that they generate $\mathcal{J}$.

If $P \in \mathcal{J}$, let us show, by induction on $\deg(P)$, that P is a combination, with coefficient in $A[X]$, of the polynomials P_{ij}.

If $\deg(P) = 0$, then $P \in \mathcal{I}_0$, and P is a combination of the P_{0j}.

Assume $\deg(P) = t > 0$. Let a be the leading coefficient of P.

If $t \leq m$, there is a decomposition $a = b_1 a_{t1} + \cdots + b_{n_t} a_{tn_t}$. The degree of the polynomial $P - (b_1 P_{t1} + \cdots + b_{n_t} P_{tn_t})$ is strictly less than t. Since this polynomial is in $\mathcal{J}$, it is a combination of the polynomials P_{ij}, by the induction hypothesis, and so is P.

If $t \geq m$, there is a decomposition $a = b_1 a_{m1} + ... + b_{n_m} a_{mn_m}$. The degree of the polynomial $P - (b_1 X^{t-m} P_{m1} + \cdots + b_{n_m} X^{t-m} P_{mn_m})$ is strictly less than t. Since this polynomial is in $\mathcal{J}$, it is a combination of the polynomials P_{ij}, by the induction hypothesis, and so is P. $\qquad\square$

Corollary 3.7 *If A is a Noetherian ring, an A-algebra of finite type is a Noetherian ring.*

Proof The ring $A[X_1, ..., X_n]$ is Noetherian and a quotient of this ring also. $\qquad\square$

3.2 Noetherian UFDs

Theorem 3.8 *A Noetherian domain A is a UFD if and only if any irreducible element of A generates a prime ideal.*

Proof It suffices to show that in a Noetherian domain A any non-zero element is a product of units and irreducible elements.

First note that if $aA = a'A$, then a is a product of units and irreducible elements if and only if a' is such a product.

We can therefore consider the set of principal ideals of the form aA where a is not a product of units and irreducible elements. If this set is not empty, let cA be a maximal element of this set. Since c is neither irreducible nor a unit, there is a decomposition $c = bb'$, where b and b' are not invertible. But, obviously, cA is strictly contained in the ideals bA and $b'A$. This shows that both b and b' are products of units and irreducible elements. Hence c also, a serious contradiction. $\qquad\square$

3.3 Primary decomposition in Noetherian rings

Definition 3.9 *A proper ideal $\mathcal{I} \subset A$ is irreducible if*

$$\mathcal{I} = \mathcal{I}_1 \cap \mathcal{I}_2 \quad \text{and} \quad \mathcal{I} \neq \mathcal{I}_1 \Rightarrow \mathcal{I} = \mathcal{I}_2.$$

Note that a prime ideal is irreducible by 1.36.

Exercises 3.10

1. Show that $\mathcal{I}$ is irreducible if and only if (0) is irreducible in $A/\mathcal{I}$.
2. Show that if $p \in \mathbb{Z}$ is prime and $n > 0$, then $p^n \mathbb{Z}$ is an irreducible ideal of $\mathbb{Z}$.

Definition 3.11 *A proper ideal $\mathcal{I} \subset A$ is primary if*

$$ab \in \mathcal{I} \text{ and } a \notin \mathcal{I} \Rightarrow b^n \in \mathcal{I} \text{ for } n \gg 0.$$

Proposition 3.12 *A proper ideal $\mathcal{I}$ is primary if and only if any zero divisor in $A/\mathcal{I}$ is nilpotent.*

Proof Note that $\mathrm{cl}(a)\,\mathrm{cl}(b) = 0 \in A/\mathcal{I} \Longleftrightarrow ab \in \mathcal{I}$. Assume $a \notin \mathcal{I}$, in other words $\mathrm{cl}(a) \neq 0$. Since $b^n \in \mathcal{I}$ for $n \gg 0$ if and only if $\mathrm{cl}(b) \in A/\mathcal{I}$ is nilpotent, we are done. $\qquad\square$

Exercises 3.13

1. Prove that a prime ideal is primary.
2. Show that if $\sqrt{\mathcal{I}}$ is maximal, $\mathcal{I}$ is primary.

Theorem 3.14 *In a Noetherian ring an irreducible ideal is primary.*

Proof It suffices to show that if (0) is irreducible, then it is primary.
Assume $ab = 0$, with $a \neq 0$. Consider the increasing sequence of ideals

$$((0) : b) \subset ((0) : b^2) \subset \dots \subset ((0) : b^n) \subset \dots$$

Since it is stationary, there exists n such that $((0) : b^n) = ((0) : b^{n+1})$.
Let us show $(0) = b^n A \cap aA$. If $b^n c = ad = x \in b^n A \cap aA$, we have $b^{n+1}c = bad = 0$, hence $c \in ((0) : b^{n+1}) = ((0) : b^n)$, and $x = b^n c = 0$. Since (0) is irreducible and $aA \neq (0)$ we have proved $b^n A = (0)$, hence $b^n = 0$. $\qquad\square$

Corollary 3.15 *In a Noetherian ring every ideal is a finite intersection of primary ideals.*

Proof If not, let E be the set of those ideals which are not a finite intersection of primary ideals. Let $\mathcal{I}$ be a maximal element of E.

Since $\mathcal{I} \in E$ it cannot be irreducible. Therefore $\mathcal{I} = \mathcal{I}_1 \cap \mathcal{I}_2$, with $\mathcal{I} \neq \mathcal{I}_1$ and $\mathcal{I} \neq \mathcal{I}_2$. Since $\mathcal{I}_1$ and $\mathcal{I}_2$ are finite intersections of primary ideals, $\mathcal{I}$ must also be and there is a contradiction. $\qquad\square$

Definition 3.16 *If $\mathcal{Q}_i$, with $i = 1, ..., n$, are primary ideals, we say that*

$$\mathcal{I} = \bigcap_1^n \mathcal{Q}_i$$

is a primary decomposition of $\mathcal{I}$.

Exercise 3.17 If $n = p_1^{r_1} \ldots p_m^{r_m}$ is a prime decomposition of the integer n, show that $n\mathbb{Z} = p_1^{r_1}\mathbb{Z} \cap \ldots \cap p_m^{r_m}\mathbb{Z}$ is a primary decomposition of $n\mathbb{Z}$ in the Noetherian ring $\mathbb{Z}$.

3.4 Radical of an ideal in a Noetherian ring

Proposition 3.18 *If $\mathcal{I}$ is an ideal in a Noetherian ring A, there exists n such that $(\sqrt{\mathcal{I}})^n \subset \mathcal{I}$.*

Proof Let $(a_1, ..., a_r)$ be a system of generators of $\sqrt{\mathcal{I}}$ and $n_1, ..., n_r$ integers such that $a_i^{n_i} \in \mathcal{I}$. Let $l_1, ..., l_r$ be positive integers such that $l_1 + \cdots + l_r \geq \sum n_i - (r - 1)$. This implies $l_i \geq n_i$ for some i and shows $a_1^{l_1} a_2^{l_2} \ldots a_r^{l_r} \in \mathcal{I}$. Consequently, $(\sqrt{\mathcal{I}})^n \subset \mathcal{I}$ for $n \geq \sum n_i - (r - 1)$. $\qquad\square$

Proposition 3.19 *If $\mathcal{I}$ is a primary ideal of a Noetherian ring, then $\sqrt{\mathcal{I}}$ is prime.*

Proof Assume $ab \in \sqrt{\mathcal{I}}$. Let n be such that $a^n b^n \in \mathcal{I}$. Then $a \notin \sqrt{\mathcal{I}}$ implies $a^n \notin \sqrt{\mathcal{I}}$, hence $a^n \notin \mathcal{I}$. Since $\mathcal{I}$ is primary, $(b^n)^m \in \mathcal{I}$, for $m \gg 0$, hence $b \in \sqrt{\mathcal{I}}$. $\qquad\square$

DANGER: An ideal whose radical is prime is not necessarily primary.

Exercise 3.20 Let $\mathcal{I} = (X) \cap (X^2, Y^2) \subset k[X, Y]$, where k is a field. Show that $\sqrt{\mathcal{I}} = (X)$ and that $\mathcal{I}$ is not primary.

Considering the definition of a primary ideal, we have:

Proposition 3.21 *Let $\mathcal{I}$ be an ideal with prime radical $\mathcal{P} = \sqrt{\mathcal{I}}$. Then $\mathcal{I}$ is primary if and only if all zero divisors modulo $\mathcal{I}$ are in $\mathcal{P}$.*

Definition 3.22 *Let $\mathcal{P}$ be a prime ideal. A primary ideal $\mathcal{I}$ whose radical is $\mathcal{P}$ is $\mathcal{P}$-primary.*

Proposition 3.23 *If $\mathcal{I}$ and $\mathcal{J}$ are both $\mathcal{P}$-primary ideals, then $\mathcal{I} \cap \mathcal{J}$ is $\mathcal{P}$-primary.*

Proof Obviously, $\mathcal{P}$ is the radical of $\mathcal{I} \cap \mathcal{J}$. Assume $ab \in \mathcal{I} \cap \mathcal{J}$. If $a \notin \mathcal{I} \cap \mathcal{J}$, then $a \notin \mathcal{I}$, for example. This implies $b^n \in \mathcal{I}$ for $n \gg 0$, hence $b \in \mathcal{P}$ and consequently b is nilpotent modulo $\mathcal{I} \cap \mathcal{J}$. $\qquad\qquad\square$

3.5 Back to primary decomposition in Noetherian rings

Definition 3.24 *Let $\mathcal{I} = \bigcap_1^n \mathcal{Q}_i$ be a primary decomposition. If $\mathcal{I} \neq \bigcap_{i \neq j} \mathcal{Q}_i$ for all j and if $\sqrt{\mathcal{Q}_i} \neq \sqrt{\mathcal{Q}_j}$ for $i \neq j$, this decomposition is called minimal.*

Using Proposition 3.23, it is clear that in a Noetherian ring all ideals have a minimal primary decomposition.

Theorem 3.25 *Let $\mathcal{I} = \bigcap_1^n \mathcal{Q}_i$ be a minimal primary decomposition of $\mathcal{I}$ (in a Noetherian ring A) and $\mathcal{P}$ a prime ideal of A. The following conditions are equivalent:*

(i) *there exists an integer i such that $\mathcal{Q}_i$ is $\mathcal{P}$-primary;*

(ii) *there exists an element $x \in A$ such that $\mathcal{P} = \mathcal{I} : x$.*

Proof Assume first $\mathcal{P} = \sqrt{\mathcal{Q}_1}$. Since the decomposition is minimal there exists $y \in \bigcap_{i>1} \mathcal{Q}_i$ such that $y \notin \mathcal{Q}_1$. Clearly $y\mathcal{Q}_1 \subset \mathcal{I}$. Since $\mathcal{P}^n \subset \mathcal{Q}_1$, for $n \gg 0$, we have $y\mathcal{P}^n \subset \mathcal{I}$, for $n \gg 0$. Let $m \geq 1$ be the smallest integer such that $y\mathcal{P}^m \subset \mathcal{I}$ and $x \in y\mathcal{P}^{m-1}$ such that $x \notin \mathcal{I}$.

It is obvious that $\mathcal{P} \subset (\mathcal{I} : x)$. Let $z \in (\mathcal{I} : x)$. Note that $y \in \bigcap_{i>1} \mathcal{Q}_i$ implies $yA \cap \mathcal{Q}_1 \subset \mathcal{I}$. Hence $x \notin \mathcal{I}$ and $x \in yA$ imply $x \notin \mathcal{Q}_1$. Since $xz \in \mathcal{I} \subset \mathcal{Q}_1$, we have $z^l \in \mathcal{Q}_1 \subset \mathcal{P}$, for $l \gg 0$, hence $z \in \mathcal{P}$ and $\mathcal{P} = \mathcal{I} : x$.

Conversely, assume $\mathcal{P} = \mathcal{I} : x$. We have $\mathcal{P} = \mathcal{I} : x = \bigcap_1^n (\mathcal{Q}_i : x)$. Since a prime ideal is irreducible we have $\mathcal{P} = \mathcal{Q}_i : x$ for some i. This implies on the one hand that $\mathcal{Q}_i \subset \mathcal{P}$. On the other hand, this shows that all elements in $\mathcal{P}$ are zero divisors modulo $\mathcal{Q}_i$, hence, by Proposition 3.21, that $\mathcal{P} = \sqrt{\mathcal{Q}_i}$. $\quad\square$

Corollary 3.26 *If $\mathcal{I} = \bigcap_1^n \mathcal{Q}_i$ and $\mathcal{I} = \bigcap_1^m \mathcal{Q}_i'$ are minimal primary decomposition of $\mathcal{I}$, then $n = m$ and there exists a permutation τ of $[1, n]$ such that $\sqrt{\mathcal{Q}_i'} = \sqrt{\mathcal{Q}_{\tau(i)}}$.*

This straightforward consequence of Theorem 3.25 allows the following definition.

Definition 3.27 *If $\mathcal{I} = \bigcap_1^n \mathcal{Q}_i$ is a minimal primary decompositions of $\mathcal{I}$, the prime ideals $\sqrt{\mathcal{Q}_i}$ are the associated prime ideals of $\mathcal{I}$ (or $A/\mathcal{I}$). We denote this finite set of prime ideals by $\mathrm{Ass}(A/\mathcal{I})$. In particular, $\mathrm{Ass}(A)$ is the set of prime ideals associated to A (or to (0)).*

Proposition 3.28 *Let $\mathcal{I}$ be a proper ideal of a Noetherian ring. An element of the ring is a zero divisor modulo $\mathcal{I}$ if and only if it is contained in a prime ideal associated to $\mathcal{I}$.*

Proof Let $\mathcal{P}$ be a prime ideal associated to $\mathcal{I}$. There exists $x \in A$ such that $\mathcal{P} = (\mathcal{I} : x)$. Then $z \in \mathcal{P} \Leftrightarrow zx \in \mathcal{I}$. Since $x \notin \mathcal{I}$, an element of $\mathcal{P}$ is a zero divisor modulo $\mathcal{I}$.

Conversely, let $z \notin \mathcal{I}$ and $y \notin \mathcal{I}$ be such that $zy \in \mathcal{I}$. There exists i such that $y \notin \mathcal{Q}_i$. Then $yz \in \mathcal{Q}_i$ implies $z^n \in \mathcal{Q}_i$, for $n >> 0$, hence $z \in \sqrt{\mathcal{Q}_i}$. □

3.6 Minimal prime ideals

Definition 3.29 *A prime ideal $\mathcal{P}$ containing an ideal $\mathcal{I}$ is a minimal prime ideal of $\mathcal{I}$ if there is no prime ideal strictly contained in $\mathcal{P}$ and containing $\mathcal{I}$.*

Theorem 3.30 *A proper ideal $\mathcal{I}$ of a Noetherian ring has only a finite number of minimal prime ideals, all associated to $\mathcal{I}$.*

Proof Let $\mathcal{P}$ be a minimal prime of $\mathcal{I}$. If $\mathcal{I} = \bigcap_1^n \mathcal{Q}_i$ is a minimal primary decomposition of $\mathcal{I}$, there exists i such that $\mathcal{Q}_i \subset \mathcal{P}$, by Proposition 1.36. This implies $\sqrt{\mathcal{Q}_i} \subset \mathcal{P}$, and the theorem. □

CAREFUL: A prime ideal associated to $\mathcal{I}$ is not necessarily a minimal prime ideal of $\mathcal{I}$.

Example 3.31 If $\mathcal{I} = (X) \cap (X^2, Y^2) \subset k[X, Y]$ (where k is a field), show that (X, Y) is associated to $\mathcal{I}$ but is not a minimal prime of $\mathcal{I}$.

Proposition 3.32 *Let $\mathcal{I} = \bigcap_1^n \mathcal{Q}_i$ be a minimal primary decomposition of $\mathcal{I}$. If $\sqrt{\mathcal{Q}_i}$ is a minimal prime ideal of $\mathcal{I}$, then $\mathcal{Q}_i = \bigcup_{s \notin \sqrt{\mathcal{Q}_i}} (\mathcal{I} : s)$.*

Proof Put $\mathcal{P}_l = \sqrt{\mathcal{Q}_l}$ for any l. Assume $a \in \mathcal{Q}_i$. If $s \in \bigcap_{j \neq i} \mathcal{Q}_j$, then $as \in \mathcal{I}$. It suffices to prove that $\bigcap_{j \neq i} \mathcal{Q}_j \not\subset \mathcal{P}_i$. If not there exists $j \neq i$ such that $\mathcal{Q}_j \subset \mathcal{P}_i$. This implies $\mathcal{P}_j \subset \mathcal{P}_i$, hence $\mathcal{P}_j = \mathcal{P}_i$ since $\mathcal{P}_i$ is minimal. This is not possible since the primary decomposition is minimal.

Conversely if $s \notin \mathcal{P}_i$, then s is not a zero divisor modulo $\mathcal{Q}_i$. Hence $as \in \mathcal{I} \subset \mathcal{Q}_i$ implies $a \in \mathcal{Q}_i$.   □

As a remarkable consequence of this last proposition, we get the uniqueness of the $\mathcal{P}$-primary component of an ideal for a minimal prime ideal $\mathcal{P}$ of this ideal.

Corollary 3.33 *Let $\mathcal{I} = \bigcap_1^n \mathcal{Q}_i = \bigcap_1^n \mathcal{Q}_i'$ be minimal primary decompositions of $\mathcal{I}$ such that $\sqrt{\mathcal{Q}_i'} = \sqrt{\mathcal{Q}_i}$. If $\sqrt{\mathcal{Q}_i'} = \sqrt{\mathcal{Q}_i}$ is a minimal prime ideal of $\mathcal{I}$ then $\mathcal{Q}_i' = \mathcal{Q}_i$.*

3.7   Noetherian modules

Let A be a ring, not necessarily Noetherian.

Definition 3.34 *An A-module M is Noetherian if it satisfies the following equivalent (see Proposition 3.1) conditions:*

(i) *any non-empty set of submodules of M contains a maximal element;*

(ii) *all increasing sequences of submodules of M are stationary.*

Proposition 3.35 *Let M be an A-module and N a submodule of M. The following conditions are equivalent:*

(i) *M is Noetherian;*

(ii) *N and M/N are Noetherian.*

Proof $(i) \Rightarrow (ii)$ is clear enough. Let us prove $(ii) \Rightarrow (i)$.

Let M_n be an increasing sequence of sub-modules of M. Then $M_n \cap N$ and $(M_n + N)/N$ are increasing sequences of sub-modules of N and M/N. We show that

$$M_n \cap N = M_{n+1} \cap N \text{ and } (M_n + N)/N = (M_{n+1} + N)/N \Rightarrow M_n = M_{n+1}.$$

If $x \in M_{n+1}$, there exists $y \in M_n$ such that $cl(x) = cl(y) \in (M_{n+1} + N)/N$. This implies $x - y \in N$, hence $x - y \in M_{n+1} \cap N = M_n \cap N \subset M_n$ and $x \in M_n$.   □

With this proposition in view, you can solve the next exercise by an obvious induction.

Exercise 3.36 Let M_i, with $i = 1, \ldots, n$, be A-modules. Show that $\bigoplus_{i=1}^n M_i$ is Noetherian if and only if M_i is Noetherian for all i.

Theorem 3.37 *An A-module is Noetherian if and only if all its submodules are finitely generated.*

We have already seen a proof of a special case of this result (Theorem 3.4): a ring is Noetherian if and only if its ideals are all of finite type.

The proof of this generalization is essentially identical. Do it.

Proposition 3.38 *A surjective endomorphism of a Noetherian module is an automorphism.*

Proof Let u be this endomorphism. Since u is surjective, u^n is surjective for $n > 0$. The submodules $\ker(u^n)$ of the module form an increasing sequence. Hence there exists $n > 0$ such that $\ker(u^n) = \ker(u^{n+1})$. Consider $x \in \ker(u)$. There exists y such that $x = u^n(y)$. But

$$u(x) = 0 \implies u^{n+1}(y) = 0 \implies u^n(y) = 0 \implies x = 0.$$

$\square$

3.8   Exercises

1. Let A be a ring and $\mathcal{I}$ and $\mathcal{J}$ be ideals of A such that $\mathcal{I} \cap \mathcal{J} = (0)$. Show that A is a Noetherian ring if and only if $A/\mathcal{I}$ and $A/\mathcal{J}$ are Noetherian rings.

2. Let M be a finitely generated A-module such that $(0) : M = (0)$. Show that if M is a Noetherian module, then A is a Noetherian ring.

3. Let A be a Noetherian ring. Show that each zero divisor is nilpotent if and only if A has only one associated prime ideal.

4. Let A be a Noetherian ring and $a \in A$. Show that if a is not contained in any minimal prime of A, then $ab = 0$ implies that b is nilpotent.

5. Let A be a Noetherian ring, $a \in A$ and $\mathcal{P} \in \text{Ass}(A)$. Assume that a is not a zero divisor and that $\text{cl}(a) \in A/\mathcal{P}$ is not a unit. Show that there exists a prime ideal $\mathcal{P}'$ with $a \in \mathcal{P}'$ and $\mathcal{P}'/aA \in \text{Ass}(A/aA)$ and such that $\mathcal{P} \subset \mathcal{P}'$. Assume that $b \in A$ is such that $\text{cl}(b) \in A/aA$ is not a zero divisor and prove that b is not a zero divisor.

6. Let A be a Noetherian ring and $a \in A$. Assume $(a^{n+1}) : a^n = (a)$ for n large enough. Show that a is not a zero divisor.

7. Let A be a Noetherian ring and $a \in A$. Show that the subring $A[aT]$ of the polynomial ring $A[T]$ is Noetherian. Show that a is not a zero divisor if and only if the A-algebra homomorphism $\pi : (A/aA)[X] \to A[aT]/aA[aT]$ defined by $\pi(X) = \mathrm{cl}(aT)$ is an isomorphism.

8. Let A be a local Noetherian ring and $\mathcal{M}$ its maximal ideal. Assume $\mathcal{M} \in \mathrm{Ass}(A)$. Show that if L is a finite rank free A-module and $e \in L$ is an element such that $ae = 0 \Rightarrow a = 0$, then there exists a free submodule L' of L such that $L = Ae \oplus L'$.

4

Artinian rings and modules

Artinian rings are Noetherian. As we will see, this is not clear from their definition. They are in a way the "smallest" Noetherian rings. The main theorem of this section states that a ring is Artinian if and only if it is Noetherian and all its prime ideals are maximal. We considered in fact the possibility of defining Artinian rings in this way and showing that they also enjoyed the descending chain condition. But we came back to the traditional introduction of Artinian rings. Not without reluctance! Our reader should keep in mind that a field is an Artinian ring and that an Artinian ring without zero divisors is a field.

I hope, once more, that you agree with the following result. If not, prove it.

Proposition 4.1 *Let E be an ordered set. The following conditions are equivalent.*

(i) *Any non-empty subset of E has a minimal element.*

(ii) *Any decreasing sequence of E is stationary.*

4.1 Artinian rings

Definition 4.2 *If the set of ideals of a ring satisfies the equivalent conditions of the preceding proposition, the ring is Artinian.*

Examples 4.3

1. Since a field has only one ideal, it is an Artinian ring.
2. More generally a ring having only finitely many ideals is Artinian.
   For example $\mathbb{Z}/n\mathbb{Z}$, with $n \geq 2$, is Artinian.

3. The ring $\mathbb{C}[X,Y]/(X^2,Y^2,XY)$ is Artinian, but it has infinitely many ideals.

To understand the last example, note first that $\mathbb{C}[X,Y]/(X^2,Y^2,XY)$ is a $\mathbb{C}$-vector space of rank 3 and that $(1, \mathrm{cl}(X), \mathrm{cl}(Y))$ is a basis of this vector space. Check next that the ideals of the ring are the subvector spaces of the rank-2 vector space generated by $\mathrm{cl}(X)$ and $\mathrm{cl}(Y)$. From this deduce that a decreasing sequence of ideals is stationary and that there are infinitely many ideals.

Since the ideals of a quotient ring $A/\mathcal{I}$ are naturally in bijection with the ideals of A containing $\mathcal{I}$, we get the following proposition.

Proposition 4.4 *If $\mathcal{I}$ is an ideal of an Artinian ring, the quotient ring $A/\mathcal{I}$ is Artinian.*

Our next result is essential, but its proof is almost obvious.

Proposition 4.5 *An Artinian domain is a field.*

Proof Let $x \neq 0$ be an element of the ring. If x is not invertible, then $x^n A$, $n > 0$, is a decreasing sequence of ideals. Since it is stationary, $x^n A = x^{n+1} A$, for $n \gg 0$. Hence there exists $a \in A$ such that $x^n = ax^{n+1}$. Since the ring is a domain, x is not a zero divisor and $1 = ax$. □

Corollary 4.6 *In an Artinian ring all prime ideals are maximal.*

Proof Let $\mathcal{P}$ be a prime ideal. The ring $A/\mathcal{P}$ is an Artinian domain, hence a field. In other words $\mathcal{P}$ is a maximal ideal. □

Proposition 4.7 *An Artinian ring has only finitely many maximal ideals.*

Proof Assume there exist distinct maximal ideals $\mathcal{M}_i$, for all $i \in \mathbb{N}$.
Since $\mathcal{M}_i \not\subset \mathcal{M}_{n+1}$ for $i < n+1$, we have $\bigcap_0^n \mathcal{M}_i \not\subset \mathcal{M}_{n+1}$. This shows that $\mathcal{I}_n = \bigcap_0^n \mathcal{M}_i$ is a strictly decreasing sequence of ideals, hence a contradiction. □

Proposition 4.8 *Let A be an Artinian ring and $\mathcal{M}_1, ..., \mathcal{M}_r$ its maximal ideals. There exist positive integers $n_1, ..., n_r$ such that*

$$(0) = \mathcal{M}_1^{n_1} \mathcal{M}_2^{n_2} ... \mathcal{M}_r^{n_r} = \bigcap_1^r \mathcal{M}_i^{n_i}.$$

Proof Since A is Artinian, there are positive integers n_i such that $\mathcal{M}_i^{n_i} = \mathcal{M}_i^{n_i+1}$. Let us show

$$(0) = \mathcal{M}_1^{n_1}\mathcal{M}_2^{n_2}...\mathcal{M}_r^{n_r}.$$

If not, let E be the set of ideals J such that $J\mathcal{M}_1^{n_1}\mathcal{M}_2^{n_2}...\mathcal{M}_r^{n_r} \neq (0)$. It is not empty since $\mathcal{M}_i \in E$ for all i.

Consider $\mathcal{I}$ a minimal element of E. Let $x \in \mathcal{I}$ be such that

$$x\mathcal{M}_1^{n_1}\mathcal{M}_2^{n_2}...\mathcal{M}_r^{n_r} \neq (0).$$

Note that this implies $\mathcal{I} = xA$. Furthermore, since

$$x\mathcal{M}_1^{n_1}\mathcal{M}_2^{n_2}...\mathcal{M}_r^{n_r} = x\mathcal{M}_1^{n_1+1}\mathcal{M}_2^{n_2+1}...\mathcal{M}_r^{n_r+r},$$

we find that the ideal $x\mathcal{M}_1\mathcal{M}_2...\mathcal{M}_r$ is also in E. This shows

$$x\mathcal{M}_1\mathcal{M}_2...\mathcal{M}_r = xA.$$

Since xA is obviously a finitely generated A-module, this last relation implies $xA = (0)$ by Nakayama's lemma, hence a contradiction.

Finally, the ideals $\mathcal{M}_i^{n_i}$ being pairwise comaximal, we have

$$\mathcal{M}_1^{n_1}\mathcal{M}_2^{n_2}...\mathcal{M}_r^{n_r} = \bigcap_1^r \mathcal{M}_i^{n_i},$$

by Lemma 1.61, and we are done. $\qquad\square$

We can now prove the main result of this chapter.

Theorem 4.9 *A ring is Artinian if and only if it is Noetherian and all its prime ideals are maximal.*

Proof If A is Noetherian, let $(0) = \bigcap_1^r \mathcal{Q}_l$ be a primary decomposition of (0). Since all prime ideals are maximal, $\sqrt{\mathcal{Q}_l}$ is a maximal ideal $\mathcal{M}_l$. But $\mathcal{M}_l^n \subset \mathcal{Q}_l$ for $n >> 0$. Hence there are positive integers n_l such that $(0) = \bigcap_1^r \mathcal{M}_l^{n_l}$. Since the ideals $\mathcal{M}_l^{n_l}$ are pairwise comaximal we get, as in the Artinian case, a relation

$$(*) \qquad\qquad (0) = \mathcal{M}_1^{n_1}\mathcal{M}_2^{n_2}...\mathcal{M}_r^{n_r}.$$

Now consider A a ring, $\mathcal{M}_i$ maximal ideals of A and n_i positive integers such that the relation $(*)$ is satisfied. We shall prove by induction on $\sum n_i$ that A is Artinian if and only if it is Noetherian.

If $\sum n_i = 1$, the ring A is a field and we are done.

If $\sum n_i > 1$, consider the ring $A' = A/\mathcal{M}_1^{n_1}\mathcal{M}_2^{n_2}...\mathcal{M}_r^{n_r-1}$. Note that the maximal ideals of A' are $\mathcal{M}_i' = \mathcal{M}_i/\mathcal{M}_1^{n_1}\mathcal{M}_2^{n_2}...\mathcal{M}_r^{n_r-1}$, for $1 \leq i \leq r$ if

$n_r > 1$ and for $1 \leq i < r$ if $n_r = 1$. Furthermore, there is, in A' a relation $(0) = \mathcal{M}_1'^{m_1} \mathcal{M}_2'^{m_2} ... \mathcal{M}_r'^{n_r - 1}$. If A is Artinian or Noetherian, so is A'. Therefore, by the induction hypothesis, A' is both Artinian and Noetherian.

Consider now $\mathcal{I}_r$ an increasing sequence of ideals of A. It induces an increasing sequence $\mathcal{I}_r A'$ of ideals of A', necessarily stationary. The kernel of the natural application $\mathcal{I}_r \to \mathcal{I}_r A'$ is $\mathcal{I}_r \cap \mathcal{M}_1^{n_1} \mathcal{M}_2^{n_2} ... \mathcal{M}_r^{n_r - 1}$. It is easy to check that $\mathcal{I}_r$ is stationary if and only if the increasing sequence $\mathcal{I}_r \cap \mathcal{M}_1^{n_1} \mathcal{M}_2^{n_2} ... \mathcal{M}_r^{n_r - 1}$ of submodules of $\mathcal{M}_1^{n_1} \mathcal{M}_2^{n_2} ... \mathcal{M}_r^{n_r - 1}$ is stationary.

Consider next a decreasing sequence $\mathcal{J}_r$ of ideals of A. The decreasing sequence $\mathcal{J}_r A'$ of ideals of A' is stationary. We see as before that $\mathcal{J}_r$ is stationary if and only if the decreasing sequence $\mathcal{J}_r \cap \mathcal{M}_1^{n_1} \mathcal{M}_2^{n_2} ... \mathcal{M}_r^{n_r - 1}$ of submodules of $\mathcal{M}_1^{n_1} \mathcal{M}_2^{n_2} ... \mathcal{M}_r^{n_r - 1}$ is stationary.

To show our theorem, it will therefore be sufficient to prove the following assertion:

In the A-module $\mathcal{M}_1^{n_1} \mathcal{M}_2^{n_2} ... \mathcal{M}_r^{n_r - 1}$ every increasing sequence of submodules is stationary if and only if every decreasing sequence of submodules is stationary.

Now since $\mathcal{M}_1^{n_1} \mathcal{M}_2^{n_2} ... \mathcal{M}_r^{n_r} = (0)$, we have $\mathcal{M}_r(\mathcal{M}_1^{n_1} \mathcal{M}_2^{n_2} ... \mathcal{M}_r^{n_r - 1}) = (0)$. This shows that the A-module structure of $\mathcal{M}_1^{n_1} \mathcal{M}_2^{n_2} ... \mathcal{M}_r^{n_r - 1}$ is in fact induced by the structure of an $A/\mathcal{M}_r$-module. However $A/\mathcal{M}_r$ is a field, hence we have that $\mathcal{M}_1^{n_1} \mathcal{M}_2^{n_2} ... \mathcal{M}_r^{n_r - 1}$ is an $A/\mathcal{M}_r$-vector space. Its A-submodules are its $A/\mathcal{M}_r$-subvector spaces.

We can therefore conclude the proof of our theorem with the following lemma.

Lemma 4.10 *Let E be a vector space on a field k. The following conditions are equivalent:*

(i) *every decreasing sequence of subvector spaces is stationary;*

(ii) *the vector space E has finite rank;*

(iii) *every increasing sequence of subvector spaces is stationary.*

Proof of the lemma Clearly (ii) implies (i) and (iii).

Conversely, assuming that E is not of finite rank, let us prove that E has non-stationary increasing and decreasing sequences of subvector spaces. Consider z_i, for $i > 0$, linearly independent elements. Let E_i be the subvector space generated by $z_1, ..., z_i$ and F_i the subvector space generated by z_j, $j \geq i$. Then (E_i) (resp. F_i) is an infinite strictly increasing (resp. strictly decreasing) sequence of subvector spaces of E. $\qquad \square$

4.2 Artinian modules

Let A be a ring, not necessarily Noetherian.

Definition 4.11 *A module M is Artinian if it satisfies the following equivalent conditions:*

(i) *any non-empty set of submodules of M contains a minimal element;*

(ii) *all decreasing sequences of submodules are stationary.*

Proposition 4.12 *Let M be an A-module and N a sub-module of M. The following conditions are equivalent:*

(i) *the module M is Artinian;*

(ii) *the modules N and M/N are Artinian.*

Proof (i) $\Rightarrow$ (ii) is clear enough. Let us prove (ii) $\Rightarrow$ (i).

Let M_n be an decreasing sequence of sub-modules of M. Then $M_n \cap N$ and $(M_n + N)/N$ are decreasing sequences of submodules of N and M/N. It is therefore enough to show

$$M_n \cap N = M_{n+1} \cap N \text{ and } (M_n + N)/N = (M_{n+1} + N)/N \implies M_n = M_{n+1}.$$

If $x \in M_n$, there exists $y \in M_{n+1}$ such that $\mathrm{cl}(x) = \mathrm{cl}(y) \in (M_n + N)/N$. This implies $x - y \in N$, hence $x - y \in M_n \cap N = M_{n+1} \cap N \subset M_{n+1}$, and $x \in M_{n+1}$. $\qquad\square$

4.3 Exercises

1. Let A be a ring and $\mathcal{I}$ and $\mathcal{J}$ be ideals of A such that $\mathcal{I} \cap \mathcal{J} = (0)$. Show that A is an Artinian ring if and only if $A/\mathcal{I}$ and $A/\mathcal{J}$ are Artinian rings.

2. Let M be a finitely generated A-module such that $(0) : M = (0)$. Show that if M is an Artinian module, then A is an Artinian ring.

3. Let A be a Noetherian ring. Assume that if $a \in A$ is neither invertible nor nilpotent, then there exists $b \in A$ such that b is not nilpotent and that $ab = 0$. Show that A is Artinian.

4. Let M be an Artinian module. Show that an injective endomorphism of M is an automorphism of M.

5. Let A be an Artinian ring, M a finitely generated A-module and u an endomorphism of M. Show that the ring $A[u]$ is Artinian.

6. Let A be a local Artinian ring. Show that if a homomorphism $u : A^n \to A^m$ is injective, then coker u is free.

7. Let A be an Artinian ring and $a \in A$. Assume that a is not invertible. Show that there exist an integer n and an element $b \in A$ such that $a^n + b$ is invertible and $a^n b = 0$.

8. Let A be a ring. Assume there exist non-zero elements $a_1, \ldots, a_n \in A$ such that $A = (a_1, \ldots, a_n)$ and that $(0) : a_i$ is a maximal ideal for each i. Show that A is a product of fields.

5

Finitely generated modules over Noetherian rings

This is both an easy and difficult chapter. Easy because studying finitely generated modules over Noetherian rings is pleasant. There is a lot to say about them, most of it useful. Difficult because we don't want to drown the reader in details. Some choices are necessary. To begin with, we study the prime ideals associated to a finitely generated modules over a Noetherian ring (they form a finite set). Next, we move to finite length modules: this notion of length is at the root of modern algebraic geometry. We'll see in the next chapter that the length is everybody's favourite additive function. Then we go back to the very classical classification of finitely generated modules over a principal ideal ring. Presenting Krull's theorem at the end of this chapter was our last difficult choice. We do not need this result this early, but now is the right time to prove it.

Theorem 5.1 *Let A be a Noetherian ring. An A-module M is Noetherian if and only if it is finitely generated.*

Proof We already know from Theorem 3.37 that this condition is necessary.

Assume now that M is generated by $(x_1, ..., x_n)$. There exists a surjective homomorphism $nA \to M$. If K is the kernel of this homomorphism, then $M \simeq (nA)/K$. By Proposition 3.35, it is sufficient to show that nA is Noetherian. Since A is a Noetherian A-module, this is a special case of Exercise 3.36. $\square$

In this chapter *all rings are Noetherian*.

5.1 Associated prime ideals

Definition 5.2 *Let M be a module on a ring A. A prime ideal $\mathcal{P}$ of A is associated to M if there exists an element $x \in M$ such that $\mathcal{P} = (0 : x)$.*

The set of all prime ideals associated to M is denoted by $\mathrm{Ass}(M)$.

Examples 5.3

1. If A is a domain, then $\mathrm{Ass}(A) = \{(0)\}$.

2. Let $\mathcal{I}$ be an ideal of A. The elements of $\mathrm{Ass}(A/\mathcal{I})$ are the prime ideals associated to $\mathcal{I}$ (yes, the language is ambiguous but we can live with it). They have been described with the primary decomposition of $\mathcal{I}$.

3. In particular, if $\mathcal{P}$ is a prime ideal of A, then $\mathrm{Ass}(A/\mathcal{P}) = \{\mathcal{P}\}$.

Proposition 5.4 *For all non-zero $x \in M$, there exists $\mathcal{P} \in \mathrm{Ass}(M)$ such that* $(0 : x) \subset \mathcal{P}$.

Proof Consider the set of all ideals $(0 : y)$, $y \in M$, $y \neq 0$. We claim that a maximal element $(0 : z)$ of this set is a prime ideal. Assume $ab \in (0 : z)$ and $a \notin (0 : z)$. We have $abz = 0$ and $az \neq 0$. Hence $b \in (0 : az)$. But clearly, $(0 : z) \subset (0 : az)$. This implies $(0 : z) = (0 : az)$, by the maximality of $(0 : z)$, and $b \in (0 : z)$.   □

Note the two following immediate corollaries.

Corollary 5.5 *If M is an A-module, then $M \neq (0)$ if and only if $\mathrm{Ass}(M) \neq \emptyset$.*

Corollary 5.6 *For $a \in A$, the multiplication map $M \xrightarrow{a} M$, $x \to ax$ is not injective if and only if there exists $\mathcal{P} \in \mathrm{Ass}(M)$ such that $a \in \mathcal{P}$.*

Proposition 5.7 *If M' is a submodule of M, then*

$$\mathrm{Ass}(M') \subset \mathrm{Ass}(M) \subset \mathrm{Ass}(M') \cup \mathrm{Ass}(M/M').$$

Proof The inclusion $\mathrm{Ass}(M') \subset \mathrm{Ass}(M)$ is obvious.

Let $\mathcal{P} \in \mathrm{Ass}(M)$ and $x \in M$ such that $\mathcal{P} = (0 : x)$. Note first that $\mathrm{Ass}(Ax) = \{\mathcal{P}\}$.

If $M' \cap Ax \neq (0)$, then $\mathrm{Ass}(M' \cap Ax) \subset \mathrm{Ass}(Ax) = \{\mathcal{P}\}$. This shows $\mathrm{Ass}(M' \cap Ax) = \{\mathcal{P}\}$. Since $\mathrm{Ass}(M' \cap Ax) \subset Ass(M')$, we have $\mathcal{P} \in \mathrm{Ass}(M')$.

If $M' \cap Ax = (0)$, consider $\mathrm{cl}(x) \in M/M'$. We have

$$a\mathrm{cl}(x) = 0 \Leftrightarrow ax \in M' \Leftrightarrow ax \in M' \cap Ax = (0) \Leftrightarrow ax = 0.$$

This shows $(0 : \mathrm{cl}(x)) = (0 : x) = \mathcal{P}$, hence $\mathcal{P} \in \mathrm{Ass}(M/M')$.   □

Corollary 5.8 *Let M be a finitely generated A-module. If $\mathcal{P}$ is a minimal prime of $(0 : M)$, then $\mathcal{P} \in \mathrm{Ass}(M)$.*

Proof Let $(x_1, ..., x_n)$ be a system of generators of M and let $\mathcal{I}_j = (0 : x_j)$. We have $(0 : M) = \cap_j \mathcal{I}_j$. Since $(0 : M) \subset \mathcal{P}$, there exists i such that $I_i \subset \mathcal{P}$. But $A/\mathcal{I}_i \simeq Ax_i$ implies $\operatorname{Ass}(A/\mathcal{I}_i) = \operatorname{Ass}(Ax_i)$. Obviously $\mathcal{P}$ is a minimal prime of $\mathcal{I}_i$. This implies $\mathcal{P} \in \operatorname{Ass}(A/\mathcal{I}_i)$, by Theorem 3.30. We find $\mathcal{P} \in \operatorname{Ass}(Ax_i) \subset \operatorname{Ass}(M)$. $\qquad\square$

Theorem 5.9 *If M is a finitely generated A-module, there exist an increasing sequence of submodules*

$$(0) = M_0 \subset M_1 \subset ... \subset M_n = M$$

and prime ideals $\mathcal{P}_i$ such that

$$M_i/M_{i-1} \simeq A/\mathcal{P}_i \text{ for } i > 0.$$

Proof Assume there exist an increasing sequence of submodules

$$(0) = M_0 \subset M_1 \subset ... \subset M_i$$

and prime ideals $\mathcal{P}_j$, $j = 1, ..., i$, such that

$$M_j/M_{j-1} \simeq A/\mathcal{P}_j \text{ for } j = 1, ..., i.$$

We show that if $M_i \neq M$, there exist a submodule M_{i+1}, with $M_i \subset M_{i+1}$, and a prime ideal $\mathcal{P}_{i+1}$ such that $M_{i+1}/M_i \simeq A/\mathcal{P}_{i+1}$.

Indeed, if $\mathcal{P} \in \operatorname{Ass}(M/M_i)$, let $\operatorname{cl}(x) \in M/M_i$ be such that $\mathcal{P} = (0 : \operatorname{cl}(x))$. Put $M_{i+1} = Ax + M_i$ and $\mathcal{P}_{i+1} = \mathcal{P}$.

We can construct in this way an increasing sequence of submodules of M. Since M is Noetherian, this sequence is stationary. In other words, there exists n such that $M_n = M$ and we are done. $\qquad\square$

CAREFUL: The prime ideals $\mathcal{P}_i$ are certainly not uniquely defined!

This last result looks very much like a technical lemma. It is much more than that and can be interpreted in the following way. Let A be a Noetherian ring and F the free group generated by all finitely generated A-modules. In F consider the equivalence relation generated by

$$M \sim M' \text{ if } M \simeq M' \quad \text{and} \quad M \sim N + M/N \text{ if } N \subset M.$$

Then our theorem states that the quotient group of F by this equivalence relation is generated by the classes of the modules $A/\mathcal{P}$, with $\mathcal{P}$ prime ideal.

Example 5.10 Consider $M = A = \mathbb{Z}$. The following sequences satisfy our theorem:

1. $(0) = M_0 \subset M_1 = A$ with $\mathcal{P}_1 = (0)$.
2. $(0) = M_0 \subset 2\mathbb{Z} = M_1 \subset M_2 = \mathbb{Z}$, with $\mathcal{P}_1 = (0)$ and $\mathcal{P}_2 = 2\mathbb{Z}$.

Corollary 5.11 *If M is a finitely generated A-module, then $\mathrm{Ass}(M)$ is finite.*

Proof By Proposition 5.7, we have

$$\mathrm{Ass}(M) \subset \mathrm{Ass}(M_1) \cup \mathrm{Ass}(M/M_1) \subset \mathrm{Ass}(M_1) \cup \mathrm{Ass}(M_2/M_1) \cup \mathrm{Ass}(M/M_2) \subset \ldots$$

$$\mathrm{Ass}(M_1/M_0) \cup \mathrm{Ass}(M_2/M_1) \cup \ldots \cup \mathrm{Ass}(M_n/M_{n-1}) = \{\mathcal{P}_1, ..., \mathcal{P}_n\}.$$

$\square$

5.2 Finite length modules

By Theorem 4.9 an Artinian ring is a Noetherian ring whose prime ideals are all maximal. Therefore, Theorem 5.9 has the following consequence.

Theorem 5.12 *Let A be an Artinian ring and M a finitely generated A-module. There exist an increasing sequence of submodules*

$$(0) = M_0 \subset M_1 \subset \ldots \subset M_n = M$$

and maximal ideals $\mathcal{M}_i$ such that $M_i/M_{i-1} \simeq A/\mathcal{M}_i$ for $i > 0$.

Proof Since all prime ideals of A are maximal, this is clear. $\square$

Definition 5.13 *Let A be a ring (not necessarily Noetherian). A non-zero A-module M is simple if M has no submodules other than (0) and M.*

Proposition 5.14 *An A-module M is simple if and only if there exists a maximal ideal $\mathcal{M}$ such that $M \simeq A/\mathcal{M}$.*

The proof is an easy exercise left to the reader.

Definition 5.15 *Let M be an A-module. If an increasing sequence $(0) = M_0 \subset M_1 \subset \ldots \subset M_n = M$ of submodules of M is such that M_i/M_{i-1} is simple for all i, it is called a composition series of length n of M.*

We have therefore seen that a finitely generated module over an Artinian ring has a composition series.

Exercise 5.16 Let $p \in \mathbb{Z}$ be a prime number and $n > 0$. Show that $M_i = p^{n-i}\mathbb{Z}/p^n\mathbb{Z}$, with $0 \le i \le n$, is a composition series of the $\mathbb{Z}$-module $\mathbb{Z}/p^n\mathbb{Z}$.

Proposition 5.17 *If an A-module M has a composition series, it is Noetherian and Artinian.*

Proof This is an immediate consequence of Proposition 3.35 and Proposition 4.12. □

Theorem 5.18 *Let A be a ring (not necessarily Noetherian) and M an A-module having a composition series.*

(i) *Any increasing or decreasing sequence of submodules of M can be extended to a composition series.*

(ii) *All composition series of M have the same length.*

Proof We shall prove the result by induction on $l_m(M)$, the length of a composition series of M of minimal length.

If $l_m(M) = 1$, then M is simple and (i) and (ii) are then obvious.

Assume $l = l_m(M) > 1$ and let $(0) = M_0 \subset M_1 \subset ... \subset M_l = M$ be a composition series of M. If N is a submodule of M, we have

$$(M_i \cap N)/(M_{i-1} \cap N) \subset (M_i/M_{i-1}).$$

This shows that $(M_i \cap N)/(M_{i-1} \cap N)$ is simple or trivial, for all i. We can therefore extract from the sequence $M_i \cap N$ a composition series of N. This proves $l_m(N) \le l$.

Furthermore, if $l_m(N) = l$ we must have

$$(M_i \cap N)/(M_{i-1} \cap N) = (M_i/M_{i-1}),$$

for all i. This implies $M_i \cap N = M_i$, for all i, by an easy induction on i, hence $N = M$.

Let now $(0) = N_0 \subset M_1, \subset ... \subset N_n = M$ be a strictly increasing sequence of submodules of M. Since N_{n-1} is a strict submodule of M, we have $l_m(N_{n-1}) < l$, hence $n \le l$. This shows (i) and (ii). □

We can now introduce the "length" of a module.

Definition 5.19 *Let A be a ring (not necessarily Noetherian). If an A-module M has a composition series of length n, we say that M has length n and we write $l_A(M) = n$. If M has no composition series, we say that M has infinite length.*

When the base ring is implicit, we sometimes write $l(M)$ for $l_A(M)$.

Theorem 5.20 *Let A be a ring (not necessarily Noetherian), M an A-module and N a submodule of M. Then*

$$l_A(M) = l_A(N) + l_A(M/N).$$

Proof Let

$$N_0 \subset N_1 \subset \dots \subset N_r$$

be an increasing sequence of submodules of N and

$$F_0 \subset F_1 \subset \dots \subset F_s$$

be an increasing sequence of submodules of M/N.

Put $M_i = N_i$ for $i \le r$ and, for $i = 1, \dots, s$, let M_{r+i} be the submodule of M, containing N, and such that $M_{r+i}/N = F_i$. We have

$$M_i/M_{i-1} = N_i/N_{i-1}, \text{ for } i \le r \text{ and } M_i/M_{i-1} = F_{i-r}/F_{i-1-r}, \text{ for } i > r.$$

The sequence

$$M_0 \subset M_1 \subset \dots \subset M_{r+s}$$

is therefore a composition series of M if and only if $(N_i)_{0 \le i \le r}$ and $(F_j)_{0 \le j \le s}$ are composition series of N and M/N.

This proves on the one hand that if $l_A(N)$ and $l_A(M/N)$ are finite, then $l_A(M) = l_A(N) + l_A(M/N)$. On the other hand it shows that if either $l_A(N)$ or $l_A(M/N)$ is not finite, neither is $l_A(M)$. $\square$

The following characterizations of Artinian rings will often be convenient.

Theorem 5.21 *Let A be a ring. The following conditions are equivalent:*

(i) *the ring A is Artinian;*

(ii) *all finitely generated A-modules have finite length;*

(iii) *the A-module A has finite length.*

Proof (i) $\Rightarrow$ (ii) by Theorem 5.12, and (iii) $\Rightarrow$ (i) is an immediate consequence of Theorem 5.18. $\square$

Corollary 5.22 *Let A be an Artinian ring. If an A-algebra B is finitely generated as an A-module, it is an Artinian ring.*

Proof Since B has finite length as an A-module, it has finite length as a B-module. $\qquad\square$

Examples 5.23

1. Let k be a field and M a k-vector space. Then $l_k(M)$ is the rank of the vector space M.

2. If M_i, with $i = 1, ..., n$, are A-modules, then $l_A(\oplus_{i=1}^n M_i) = \sum_{i=1}^n l_A(M_i)$.

3. If A is an Artinian ring and M a free A-module of rank n, then $l_A(M) = nl_A(A)$.

Proposition 5.24 *Let K be a field and $k \subset K$ a subfield such that K is a finite rank k-vector space. If E is a finite rank K-vector space, then E is a finite rank k-vector space for the induced structure and we have*

$$l_k(E) = \mathrm{rk}_k(E) = \mathrm{rk}_k(K)\mathrm{rk}_K(E) = l_k(K)l_K(E).$$

Proof Let r be the rank of E as a K-vector space. There exists an isomorphism of K-vector spaces $E \simeq K^r$. Since this is an isomorphism of k-vector spaces, we are done. $\qquad\square$

Exercise 5.25 Let k be a field and A a k-algebra such that for all maximal ideals $\mathcal{M}$ of A, the natural application $k \to A/\mathcal{M}$ is an isomorphism. Note that an A-module M has an induced structure of a k-vector space. Show that the A-module M has finite length if and only if it has finite rank as a k-vector space and that $l_A(M) = \mathrm{rk}_k(M)$.

5.3 Finitely generated modules over principal ideal rings

Theorem 5.26 *Let A be a principal ideal ring, L a free A-module of rank n and M a submodule of L. Then:*

(i) *the module M is free;*

(ii) *there exist a basis (e_i) of L, a positive integer $m \le n$ and elements $a_i \in A$, with $1 \le i \le m$, such that*
 (a) $a_{i+1} \in a_i A$, for all i,
 (b) $(a_1 e_1, a_2 e_2, ..., a_m e_m)$ is a basis for M.

Proof We shall first prove (i), by an induction on the maximal number of linearly independent elements in M. If this number is zero, then $M = (0)$ and we are done. Assume $M \neq (0)$.

If $g \in \mathrm{Hom}_A(L, A)$, then $g(M)$ is an ideal of A. Since A is principal, we can consider $aA = u(M)$, a submodule of A, maximal among the $g(M)$, $g \in \mathrm{Hom}_A(L, A)$ and choose $e' \in M$ such that $u(e') = a$. Note that if $(f_1, ..., f_n)$ is a basis of L, there are linear forms v_i on L, such that $x = \sum_1^n v_i(x) f_i$ for all $x \in L$. Since $M \neq (0)$, there exists i such that $v_i(M) \neq (0)$. As a consequence $a \neq 0$, hence $e' \neq 0$.

Next we show that $v(e') \in aA$, for all $v \in \mathrm{Hom}_A(L, A)$. We can assume $v(e') \neq 0$. Consider $a' = \gcd(a, v(e'))$. We have

$$a' = ba + cv(e') = bu(e') + cv(e') = (bu + cv)(e'),$$

hence $a \in (bu + cv)(M)$ and $u(M) \subset (bu + cv)(M)$. From the maximality of $aA = u(M)$, we get $aA = u(M) = (bu + cv)(M)$. This implies $a' \in aA$, hence $v(e') \in aA$.

From this we deduce that there is an $e \in L$ such that $e' = ae$. Indeed, if $e' = \sum_1^n a_i f_i$, we have just proved that $a_i = v_i(e') \in aA$ for all i. So $e = \sum_1^n (a_i/a) f_i$ is our element. We note that $u(e) = 1$.

The next step is to prove

$$L = Ae \oplus \ker u \quad \text{and} \quad M = Ae' \oplus (\ker u \cap M).$$

That $Ae \cap \ker u = (0)$ is obvious. Furthermore, if $x \in L$, we have $x = u(x)e + (x - u(x)e) \in Ae + \ker u$.

Clearly, $Ae' \cap (\ker u \cap M) = (0)$ is also obvious. If $z \in M$, there is a $b \in A$ such that $u(z) = ba$. This shows $z = be' + (z - be') \in Ae' + (\ker u \cap M)$.

From the decomposition $M = Ae' \oplus (\ker u \cap M)$, we deduce that the maximal number of linearly independent elements in $(\ker u \cap M)$ is strictly smaller than the corresponding number for M. By the induction hypothesis $(\ker u \cap M)$ is free, hence so is M and we have proved (i).

We now prove (ii) by induction on $n = \mathrm{rk}(L)$.

By (i), $\ker u$ is free. Since $\mathrm{rk}(\ker u) = n - 1$, we can apply our induction hypothesis to the submodule $(\ker u \cap M)$ of $\ker u$.

There exist a basis $(e_2, ..., e_n)$ of $\ker u$, an integer $m \leq n$ and elements $a_i \in A$, $2 \leq i \leq m$, such that $a_{i+1} \in a_i A$, for all $i \geq 2$, and that $(a_2 e_2, ..., a_m e_m)$ is a basis of $(\ker u \cap M)$. Since $M = Aae \oplus (\ker u \cap M)$, if we denote $e_1 = e$ and $a_1 = a$, we must show $a_2 \in a_1 A$.

Let $v \in \mathrm{Hom}_A(L, A)$ be the linear form defined by $v(e_i) = 1$ for all i. We have $v(a_1 e_1) = a_1$, hence $a_1 A \subset v(M)$. Since $a_1 A = aA$, this implies $a_1 A = v(M)$. From this we deduce $a_2 = v(a_2 e_2) \in v(M) = a_1 A$ and the theorem is proved. $\square$

Corollary 5.27 *Let E be a finitely generated module on a principal ideal ring A. There exist integers m and r and non-invertible elements $a_1, ..., a_m \in A$ such that:*

(i) $a_{i+1} \in a_i A$ for $i \geq 1$;

(ii) $E \simeq (\bigoplus_1^m A/a_i A) \oplus rA$.

Furthermore m, r and the ideals $a_i A$ are uniquely defined by M.

Proof Let $x_1, ..., x_n \in E$ be elements generating E. If $(f_1, ..., f_n)$ is the canonical basis of nA, consider the surjective homomorphism $v : nA \to E$ defined by $v(f_i) = x_i$.

We consider the submodule $\ker v$ of nA and apply the theorem.

There is a basis $(e_1, ..., e_n)$ of nA and elements $a_1, ..., a_m \in A$, with $m \leq n$, and $a_{i+1} \in a_i A$ for $i \geq 1$, such that $(a_1 e_1, ..., a_m e_m)$ is a basis for $\ker v$.

By Exercise 2.33 (2), there is an isomorphism

$$(\bigoplus_1^m A/a_i A) \oplus (n-m)A \simeq E.$$

To complete the proof of the corollary, let us introduce the torsion submodule of a module on a domain.

Proposition 5.28 *If A is a domain and N an A-module, then the set $T(N)$ of all $x \in N$ such that $(0) : x \neq (0)$ is a submodule of N.*

Proof Let x and y be elements of $T(N)$. If $ax = 0$ and $by = 0$, then $ab(cx + dy) = 0$ for all $c, d \in A$. $\qquad\square$

Definition 5.29 *The module $T(N)$ is the torsion submodule of N.*

Consider now, as in the theorem, two decompositions of M:

$$M \simeq (\bigoplus_1^m A/a_i A) \oplus rA \simeq (\bigoplus_1^{m'} A/a'_i A) \oplus r'A.$$

They induce isomorphisms

$$T(M) \simeq \bigoplus_1^m A/a_i A \simeq \bigoplus_1^{m'} A/a'_i A,$$

and

$$M/T(M) \simeq rA \simeq r'A.$$

This last relation shows $r = r'$.

Lemma 5.30 *A finitely generated torsion module over a principal ideal ring has finite length.*

Proof Since such a module is isomorphic to a module of the form $\bigoplus_1^m A/a_iA$, it is sufficient to prove that if $a \in A$ is a non-zero element, then A/aA is of finite length. Note that the A-submodules of A/aA are the ideals of the quotient ring A/aA. Since the prime ideals of this ring are in natural bijection with the prime ideal of A containing a, they are all maximal. As a consequence A/aA is an Artinian ring, hence it has finite length as an A/aA-module and as an A-module as well.   □

We can now go back to the *proof of Corollary 5.27*

Let $\mathcal{M}$ be a maximal ideal of A such that $a_1 \in \mathcal{M}$. Since A is principal, there exists $a \in A$ such that $\mathcal{M} = aA$. If $b \in A$, we have

$$(A/bA)/a(A/bA) \simeq A/(aA + bA).$$

Since $a_i \in aA$ for all i, this implies

$$(A/a_iA)/a(A/a_iA) \simeq A/aA, \text{ and}$$

$$T(M)/aT(M) \simeq \bigoplus_1^m A/aA \simeq \bigoplus_1^{m'} A/(aA + a_i'A).$$

Since $A/(aA + a_i'A) = (0) \iff a_i' \notin aA$, this proves $m' \geq m$, hence $m' = m$, and $a_i' \in aA$ for all i.

We now prove $a_iA = a_i'A$ by induction on $\sum_1^m l_A(A/a_iA)$.

If $m = 0$, we are done. If $m > 0$, we have $l_A(aT(M)) < l_A(T(M))$. In this case the isomorphisms

$$aT(M) \simeq \bigoplus_1^m aA/a_iA \simeq \bigoplus_1^m aA/a_i'A \simeq \bigoplus_1^m A/(a_i/a)A \simeq \bigoplus_1^m A/(a_i'/a)A.$$

imply, by the induction hypothesis,

$$(a_i/a)A = (a_i'/a)A, \text{ hence } a_iA = a_i'A, \text{ for all } i.$$

The Corollary is proved.   □

Exercise 5.31 Let A be a principal ideal ring, L a finitely generated free A-module and u an injective endomorphism of L.

1. Show that coker (u) is an A-module of finite length.
2. Prove $l_A(\text{coker}(u)) = l_A(A/\det(u))$.

5.4 The Artin–Rees lemma and Krull's theorem

The Artin–Rees lemma is an apparently technical result whose proof is rather elementary. Although the beginner will not perceive its importance immediately, it has proved to be particularly useful (as in the case of Nakayama's lemma for example). Krull's theorem, which we will need often in the sequel, is an immediate consequence of this "lemma".

Definition 5.32 *Let $\mathcal{I}$ an ideal of a ring A. The subring $(\bigoplus_{n \geq 0} \mathcal{I}^n T^n)$ of the polynomial ring $A[T] = \bigoplus_{n \geq 0} A T^n$ is the Rees ring of A with respect to $\mathcal{I}$.*

Proposition 5.33 *If A is a Noetherian ring, the Rees ring of A with respect to an ideal $\mathcal{I}$ is Noetherian.*

Proof Let $(a_1, ..., a_r)$ be a system of generators of $\mathcal{I}$. The homomorphism of A-algebras

$$f : A[X_1, ..., X_r] \to \bigoplus_{n \geq 0} \mathcal{I}^n T^n, \quad f(X_i) = a_i T$$

is obviously surjective, hence the Rees ring is a finitely generated A-algebra. $\qquad\square$

Lemma 5.34 *(Artin–Rees lemma) Let $\mathcal{I}$ be an ideal of a Noetherian ring A and M a finitely generated A-module. If N is a submodule of M, there exists an integer m such that*

$$\mathcal{I}^{n+m} M \cap N = \mathcal{I}^n (\mathcal{I}^m M \cap N), \text{ for all } n \geq 0.$$

Proof Consider $MT^n = \{xT^n, \ x \in M\}$. Clearly, it has the structure of an A-module isomorphic to M.

The direct sum $\bigoplus_{n \geq 0} MT^n$ has an obvious structure of an $A[T]$-module, hence an induced structure of a module over the Rees ring $(\bigoplus_{n \geq 0} \mathcal{I}^n T^n)$. Note furthermore that $(\bigoplus_{n \geq 0} \mathcal{I}^n MT^n)$ is a $(\bigoplus_{n \geq 0} \mathcal{I}^n T^n)$-submodule of this module.

Now if $(x_1, ..., x_s)$ is a finite system of generators of M as an A-module, then $(x_1 T^0, ..., x_s T^0)$ is a system of generators of $(\bigoplus_{n \geq 0} \mathcal{I}^n T^n M)$ as a module over the Rees ring $(\bigoplus_{n \geq 0} \mathcal{I}^n T^n)$. Hence this module is finitely generated.

Put $N_n = \mathcal{I}^n M \cap N$. Clearly, we have $\mathcal{I}^k N_n \subset N_{n+k}$ for $k \geq 0$. This shows that $(\bigoplus_{n \geq 0} N_n T^n)$ is a $(\bigoplus_{n \geq 0} \mathcal{I}^n T^n)$-submodule of $(\bigoplus_{n \geq 0} \mathcal{I}^n MT^n)$. Consequently, $(\bigoplus_{n \geq 0} N_n T^n)$ is a finitely generated module over the Rees ring $\bigoplus_{n \geq 0} \mathcal{I}^n T^n$.

Consider $z_1, ..., z_l \in \bigoplus_{n \geq 0} N_n T^n$, a system of generators of this module. Let m be an integer such that $z_i \in (\bigoplus_{n=0}^{m} N_n T^n)$, for $i = 1, ..., l$. We have

$$\mathcal{I}^k T^k N_m T^m = N_{m+k} T^{m+k}, \text{ in other words } \mathcal{I}^k(\mathcal{I}^m M \cap N) = \mathcal{I}^{n+m} M \cap N,$$

and the lemma is proved. $\qquad\square$

Corollary 5.35 *(Krull's theorem) Let A be a Noetherian ring and M a finitely generated A-module. If $\mathcal{I}$ is an ideal of A contained in the Jacobson radical of the ring, then*

$$\bigcap_n \mathcal{I}^n M = (0).$$

Proof Put $E = \bigcap_m \mathcal{I}^n M$. We have $\mathcal{I}^n M \cap E = E$ for all $n \geq 0$. Hence by the Artin–Rees lemma $\mathcal{I}E = E$. Applying Nakayama's lemma, we have proved $E = (0)$.   □

5.5   Exercises

1. Let A be a Noetherian ring, M a finitely generated A-module and $a \in JR(A)$ an element such that $x \in M$ and $ax = 0$ imply $x = 0$ (we say that a is regular in M). If $\mathcal{P} \in \mathrm{Ass}(M)$, show that there exists a prime ideal $\mathcal{P}'$ containing a, such that $\mathcal{P}' \in \mathrm{Ass}(M/aM)$ and such that $\mathcal{P} \subset \mathcal{P}'$. Show that if $b \in A$ is regular in M/aM, then b is regular in M.

2. Let A be a Noetherian ring, M a finitely generated A-module and $\mathcal{P}$ a prime ideal such that $(0) : M \subset \mathcal{P}$. Show that there exists a submodule M' of M such that $\mathcal{P} \in \mathrm{Ass}(M/M')$.

3. Let A be a Noetherian ring and N a finitely generated A-module. Show that there exists a homomorphism of free modules $f : nA \to mA$ such that $N = \mathrm{coker}\, f$. Such a homomorphism is called a finite presentation of N. If $(e_k)_{1 \leq k \leq n}$ and $(f_l)_{1 \leq l \leq m}$ are the canonical bases of nA and mA, put $f(e_j) = \sum_i a_{ij} f_i$ and consider the matrix $M = (a_{ij})$ of f. Define $\mathcal{I}_r$ to be the ideal generated by the $(m-r)$-minors of M and show that the ideals $\mathcal{I}_r$ do not depend on the presentation f but only on N.

   The ideal $\mathcal{I}_r$ is called the rth Fitting ideal of N and often denoted by $F_r(N)$.

4. Let A be a Noetherian ring and N a finitely generated A-module. If N' is a submodule of N, show that the Fitting ideals satisfy the relations

   $$F_r(N) \subset F_r(N/N') \quad \text{and} \quad F_r(N) \subset F_r(N').$$

5. Let A be a Noetherian domain such that each non-zero prime ideal is maximal. Show that for every finitely generated A-module M, the torsion submodule $T(M)$ of M has finite length.

6. Let R be a principal ideal ring and M a finitely generated R-module. If M' is a submodule of M, show $l_R(T(M)) \leq l_R(T(M')) + l_R(T(M/M'))$.

   Show that if equality holds, then

$$M/T(M) \simeq M'/T(M') \oplus (M/M')/T(M/M').$$

7. Let A be a Noetherian ring and $a \in \mathrm{JR}(A)$ an element such that aA is a non-minimal prime ideal. Show that A is a domain.

8. Let A be a local Noetherian ring. Assume that the maximal ideal $\mathcal{M}$ of A is in $\mathrm{Ass}(A)$. Show that if a homomorphism $u : nA \to mA$ is injective, then coker u is free.

6

A first contact with homological algebra

What is homological algebra? Where does it begin? Homomorphism modules are introduced with linear algebra and often studied in commutative algebra. They certainly are "homological" objects of the same nature as tensor products, which are usually introduced much later, with homological algebra. Let's take a break and present a set of elementary algebraic methods having a common flavour. This common flavour shall be our first contact with homological algebra. Reader, if you are in a hurry be sure to understand the second section of this chapter; if not, try the other sections also.

We cannot be really consistent without abelian categories, but we have no desire to say more than necessary in our first section. Our second section deals with exact sequences and additive functions. This is important and allows us to see some previous results in a new light. In our third section we come back to homomorphism modules and present them with tensor products. Unfortunately their importance will not appear immediately to our reader, who should trust us nevertheless. In our last sections we study duality on artinian rings and introduce Gorenstein artinian rings. We hope that you will take pleasure in reading these last two sections.

6.1   Some abelian categories

A "category" consists of objects and arrows (or morphisms). The set of arrows from an object E to an object F is denoted by $\mathrm{Hom}(E, F)$. The arrows compose in a natural associative way and for each object E, there is an identity $1_E \in \mathrm{Hom}(E, E)$ which is an identity element for the composition.

The category whose objects are abelian groups and whose arrows are group homomorphisms is of common use. We are interested in some of its subcategories. More precisely, we require that all objects in our categories are abelian groups, that the direct sum of two objects is an object, and that the set $\mathrm{Hom}(E, F)$ is a subgroup of the group of homomorphisms from E to F. Furthermore, each arrow has a kernel and a cokernel in the category, and if both are zero the map is an isomorphism.

Let A be a ring. There is an "abelian category" whose objects are the A-modules and whose arrows are the homomorphisms of A-modules. Note that the kernel and the cokernel of an arrow are objects of the category.

There is also an "abelian subcategory" whose objects are the A-modules of finite length and whose arrows are the homomorphisms of such A-modules. Note once more that the kernel and the cokernel of a homomorphism between finite length A-modules have finite length.

If A is Noetherian, there is another "abelian subcategory" whose objects are the finitely generated A-modules and whose arrows are the homomorphisms of finitely generated A-modules. If the ring is not Noetherian we run into difficulties because the kernel of a homomorphism of finitely generated modules is not necessarily finitely generated.

A functor F from an abelian category C to another abelian category, C', transforms an object M of C into an object $F(M)$ of C' and an arrow of C into an arrow of C'. The functor is called covariant if it preserves the directions, in other words if the transform of an arrow $f : M \to N$ is an arrow $F(f) : F(M) \to F(N)$. It is called contravariant if it inverts these directions, in other words if the transform of an arrow $f : M \to N$ is an arrow $F(f) : F(N) \to F(M)$.

A functor has also the following properties

$$F((0)) = (0), \quad F(\mathrm{Id}_M) = \mathrm{Id}_{F(M)},$$

$$F(g \circ f) = F(g) \circ F(f) \quad \text{for } F \text{ covariant} \quad \text{and}$$

$$F(g \circ f) = F(f) \circ F(g) \quad \text{for } F \text{ contravariant.}$$

Let F be a covariant functor. If for each injective (resp. surjective) homomorphism f the homomorphism $F(f)$ is injective (resp. surjective), we say that F is left (resp. right) exact. If F is both left and right exact, we say that F is exact.

Let F be a contravariant functor. If for each surjective (resp. injective) homomorphism f the homomorphism $F(f)$ is injective (resp. surjective), we say that F is left (resp. right) exact.

We do not intend to be more precise for the time being.

Examples 6.1 Let A be a ring.

1. For all A-modules N, put $F(N) = \mathrm{Hom}_A(N, A) = N^\vee$. If $f : N \to M$ is a homomorphism, denote by $F(f)$ the transposed homomorphism $f^\vee : M^\vee \to N^\vee$. Then F is a contravariant functor from the category of A-modules to itself. If f is surjective, then $f^\vee$ is injective, in other words F is left exact.

2. Let B be an A-algebra. For all A-modules N, put $G(N) = \mathrm{Hom}_A(B, N)$. Note that $G(N)$ has a natural structure of a B-module: if $b \in B$ and $h \in \mathrm{Hom}_A(B, N)$, define bh by $(bh)(c) = h(bc)$ for all $c \in B$. If $f : N \to M$ is a homomorphism of A-modules, the natural application

$$G(f) : \mathrm{Hom}_A(B, N) \to \mathrm{Hom}_A(B, M), \quad G(f)(h) = f \circ h$$

is B-linear. Clearly, G is a covariant functor from the category of A-modules to the category of B-modules. Checking that G is left exact is once more straightforward.

6.2 Exact sequences

Definition 6.2 *Let* $f : M' \to M$ *and* $g : M \to M''$ *be homomorphisms of A-modules.*

(i) *If* $g \circ f = 0$, *i.e. if* $f(M') \subset \ker g$, *then*

$$M' \xrightarrow{f} M \xrightarrow{g} M''$$

is a complex of A-modules. The homology module of this complex is $H = \ker g / f(M')$.

(ii) *If furthermore* $f(M') = \ker g$, *in other words if the complex has trivial homology, this complex is called an exact sequence.*

Examples 6.3 Let $M \xrightarrow{f} N$ be a homomorphism of A-modules.

1. Since $f(0) = 0$, there is a complex $0 \to M \xrightarrow{f} N$. This complex is an exact sequence if and only if f is injective.

2. The complex $M \xrightarrow{f} N \to 0$ is an exact sequence if and only if f is surjective.

3. There is an exact sequence

$$0 \to \ker f \xrightarrow{i} M \xrightarrow{f} N \xrightarrow{cl} \mathrm{coker}\, f \to 0,$$

where i is the natural inclusion and cl : $N \to N/f(M)$ the natural surjection.

4. A commutative diagram

$$\begin{array}{ccc} M & \xrightarrow{f} & N \\ \downarrow & & \downarrow \\ M' & \xrightarrow{f'} & N' \end{array}$$

induces the following commutative diagram, whose two lines are exact:

$$0 \rightarrow \ker f \rightarrow M \xrightarrow{f} N \rightarrow \operatorname{coker} f \rightarrow 0$$
$$\downarrow \qquad \downarrow \qquad \downarrow \qquad \downarrow$$
$$0 \rightarrow \ker f' \rightarrow M' \xrightarrow{f'} N' \rightarrow \operatorname{coker} f' \rightarrow 0.$$

Definition 6.4 *Let* $0 \rightarrow M' \xrightarrow{f} M \xrightarrow{g} M'' \rightarrow 0$ *be an exact sequence of A-modules. If the module* $f(M') = \ker g$ *is a direct factor of M, we say that the sequence splits.*

Definition 6.5 *A function* λ *defined on the category of A-modules (resp. finite length A-modules, finitely generated A-modules if A is Noetherian), with value in an abelian group G, is additive if, for all exact sequences* $0 \rightarrow M' \xrightarrow{f} M \xrightarrow{g} M'' \rightarrow 0$, *we have*

$$\lambda(M) = \lambda(M') + \lambda(M'').$$

Examples 6.6

1. The rank, defined in the category of finite rank k-vector spaces (where k is a field), with value in $\mathbb{Z}$, is an additive function.

2. The length, defined in the category of finite length A-modules, with value in $\mathbb{Z}$, is an additive function.

Exercise 6.7 Let A be a principal ideal ring. If M is a finitely generated A-module, we recall that $M/T(M)$ (where $T(M)$ is the torsion submodule of M) is a free A-module. We define $\operatorname{rk}_A(M) = \operatorname{rk}(M/T(M))$.

Show that $\operatorname{rk}_A(*)$ is an additive function on the category of finitely generated A-modules.

The following property of additive functions is practically contained in their definition. It is important. Keep it in mind; we will use it in section 4.

Proposition 6.8 *Let* $0 \xrightarrow{f_{r+1}} M_r \xrightarrow{f_r} M_{r-1} \rightarrow ... \xrightarrow{f_2} M_1 \xrightarrow{f_1} M_0 \xrightarrow{f_0} 0$ *be a complex of A-modules (resp. finite length A-modules, finitely generated A-modules if A is Noetherian). If* λ *is an additive function defined in the corresponding category, then*

$$\sum_i (-1)^i \lambda(M_i) = \sum_i (-1)^i \lambda(\ker(f_i)/f_{i+1}(M_{i+1})).$$

It may be surprising for the reader, but the next result is truly important! The fundamental piece of information is that the commutative diagram presented induces a "canonical map" $\ker g'' \to \operatorname{coker} g'$, a "connection homomorphism". That this homomorphism fits nicely in a long exact sequence, makes it even more useful.

Theorem 6.9 *(The snake lemma)*
    Let

$$\begin{array}{ccccccc}
M' & \xrightarrow{t} & M & \xrightarrow{u} & M'' & \to & 0 \\
{\scriptstyle g'}\downarrow & & {\scriptstyle g}\downarrow & & {\scriptstyle g''}\downarrow & & \\
0 & \to & N' & \xrightarrow{v} & N & \xrightarrow{w} & N''
\end{array}$$

be a commutative diagram whose lines are exact sequences.
    There exists a long exact sequence

$$\ker g' \to \ker g \to \ker g'' \to \operatorname{coker} g' \to \operatorname{coker} g \to \operatorname{coker} g''$$

whose homomorphisms commute with the homomorphisms of the diagram.
    If furthermore t is injective (resp. w is surjective), then $\ker g' \to \ker g$ *is injective (resp.* $\operatorname{coker} g \to \operatorname{coker} g''$ *is surjective).*

Proof By Example 6.3 (4), we can enlarge our commutative diagram as follows

$$\begin{array}{ccccccc}
0 & & 0 & & 0 & & \\
\downarrow & & \downarrow & & \downarrow & & \\
\ker g' & \to & \ker g & \to & \ker g'' & & \\
\downarrow & & \downarrow & & \downarrow & & \\
M' & \xrightarrow{t} & M & \xrightarrow{u} & M'' & \to & 0 \\
\downarrow{\scriptstyle g'} & & \downarrow{\scriptstyle g} & & {}'\downarrow{\scriptstyle g''} & & \\
0 \to \quad N' & \xrightarrow{v} & N & \xrightarrow{w} & N'' & & \\
\downarrow & & \downarrow & & \downarrow & & \\
\operatorname{coker} g' & \to & \operatorname{coker} g & \to & \operatorname{coker} g'' & & \\
\downarrow & & \downarrow & & \downarrow & & \\
0 & & 0 & & 0 & &
\end{array}$$

where the rows are exact sequences.
    Proving that the complexes

$$\ker g' \to \ker g \to \ker g'' \quad \text{and} \quad \operatorname{coker} g' \to \operatorname{coker} g \to \operatorname{coker} g''$$

are exact is straightforward.
    The only difficulty is to construct the "connection homomorphism"

$$\ker g'' \xrightarrow{c} \operatorname{coker} g'.$$

This is "diagram chasing". To this end consider $x'' \in \ker g''$. Choose any $x \in M$ such that $x'' = u(x)$ and define $y = g(x)$. Note then that

$$w(y) = g''(x'') = 0.$$

Hence there exists $y' \in N'$ such that $y = v(y')$. It is easy to check that the class $\overline{y'} \in N'/g'(M') = \operatorname{coker} g'$ of y' does not depend on the arbitrary choice made. We define $c(x'') = \overline{y'} \in \operatorname{coker} g'$.

The proof of the lemma may be completed easily. We show for example, that

$$\ker g \to \ker g'' \xrightarrow{c} \operatorname{coker} g'$$

is an exact sequence.

If $x'' = u(x)$ with $x \in \ker g$, then $y = 0$ and $c(x'') = \overline{y'} = 0$.

If $c(x'') = \overline{y'} = 0$, there exists $x' \in M'$ such that $y' = g'(x')$. In this case, we have $x'' = u(x - t(x'))$ where $(x - t(x')) \in \ker g$.   □

As an immediate consequence, we get the next corollary:

Corollary 6.10

(i) *If g' and g'' are injective, so is g.*

(ii) *If g is injective and g' surjective, then g'' is injective.*

(iii) *If g'' is injective and g surjective, then g' is surjective.*

(iv) *If g' and g'' are surjective, so is g.*

Corollary 6.11 *If M_1 and M_2 are submodules of M, there is a natural exact sequence*

$$0 \to M/(M_1 \cap M_2) \to M/M_1 \oplus M/M_2 \to M/(M_1 + M_2) \to 0.$$

Proof Consider the following commutative diagram:

$$
\begin{array}{ccccccccc}
0 & \to & M_1 \cap M_2 & \to & M_1 \oplus M_2 & \to & M_1 + M_2 & \to & 0 \\
 & & \downarrow & & \downarrow & & \downarrow & & \\
0 & \to & M & \xrightarrow{f} & M \oplus M & \xrightarrow{g} & M & \to & 0,
\end{array}
$$

where $f(x) = (x, x)$ and $g(x, y) = x - y$ and the verticale arrows are the natural inclusions. It is clear that the two lines are exact. Our corollary is the snake lemma in this case.   □

Let us focus for a while on the following special case. If $\mathcal{I}$ and $\mathcal{J}$ are ideals of a ring A, there is a natural exact sequence

$$0 \to A/(\mathcal{I} \cap \mathcal{J}) \to A/\mathcal{I} \oplus A/\mathcal{J} \to A/(\mathcal{I} + \mathcal{J}) \to 0.$$

This proves that $\mathcal{I}$ and $\mathcal{J}$ are comaximal if and only if the natural map

$$A/(\mathcal{I} \cap \mathcal{J}) \to A/\mathcal{I} \oplus A/\mathcal{J}$$

is an isomorphism.

An easy induction on n proves that $(\mathcal{I}_i)_{1 \le i \le n}$ are pairwise comaximal ideals if and only if the natural application

$$A/(\cap_1^n \mathcal{I}_i) \to \bigoplus_1^n A/\mathcal{I}_i$$

is an isomorphism. This is our Theorem 1.62.

6.3 Tensor products and homomorphism modules

Although we will not use it immediately, it seems a good time for a first contact with the tensor product, $M \otimes_A N$, of two A-modules M and N. In our discussion, we reconsider the homomorphism module $\mathrm{Hom}_A(M, N)$ that incidentally we met earlier.

Consider two A-modules M and N and their product $M \times N$. We denote by $\bigoplus_{(x,y) \in M \times N} A(x, y)$ the free A-module with a basis indexed by $M \times N$. Next we consider the submodule R of $\bigoplus_{(x,y) \in M \times N} A(x, y)$ generated by the elements

$$((x_1 + x_2, y) - (x_1, y) - (x_2, y)), \quad ((x, y_1 + y_2) - (x, y_1) - (x, y_2)),$$

$$((ax, y) - a(x, y)) \quad \text{and} \quad ((x, ay) - a(x, y)).$$

Definition 6.12

(i) *The tensor product $M \otimes_A N$ is the A-module $\bigoplus_{(x,y) \in M \times N} A(x, y)/R$. The class of (x, y) is denoted by $x \otimes_A y$.*

(ii) *If $N \xrightarrow{f} N'$ is a homomorphism, then the homomorphism*

$$M \otimes_A f : (M \otimes_A N) \to (M \otimes_A N')$$

is defined by

$$(M \otimes_A f)(x \otimes_A y) = (x \otimes_A f(y)).$$

This apparently monstrous object is in fact natural and convenient. Its positive nature appears already in the following assertion.

Theorem 6.13

(i) *The natural map* $b : M \times N \to M \otimes_A N$, $\quad b(x, y) = x \otimes_A y$ *is A-bilinear.*

(ii) *For any A-bilinear application* $f : M \times N \to P$ *(where P is an A-module), there exists a unique factorization* $f = u \circ b$, *through a homomorphism* $u : M \otimes_A N \to P$ *of A-modules.*

The proof of this theorem is straightforward and left to the reader.

Proposition 6.14 *There are natural isomorphisms:*

(i) $M \otimes_A N \simeq N \otimes_A M$,

(ii) $A \otimes_A M \simeq M$,

(iii) $(M \otimes_A N) \otimes_A P \simeq M \otimes_A (N \otimes_A P)$,

(iv) $(M \oplus N) \otimes_A P \simeq (M \otimes_A P) \oplus (N \otimes_A P)$.

These isomorphisms are clear enough. The following is a bit more intricate.

Proposition 6.15 *The natural homomorphism*

$$\operatorname{Hom}_A(M \otimes_A N, P) \to \operatorname{Hom}_A(M, \operatorname{Hom}_A(N, P))$$

is an isomorphism.

Proof If $f \in \operatorname{Hom}_A(M \otimes_A N, P)$, then clearly $f(x \otimes_A \,.\,) \in \operatorname{Hom}_A(N, P)$. Our map is defined and obviously injective. Now if $g \in \operatorname{Hom}_A(M, \operatorname{Hom}_A(N, P))$, note that

$$M \times N \to P, \quad (x, y) \to g(x)(y)$$

is an A-bilinear application. To conclude, we use Theorem 6.13 (ii).   □

Next we want to study the functors associated to the tensor product and the homomorphism modules.

Proposition 6.16 *Let* $0 \to M' \xrightarrow{f} M \xrightarrow{g} M'' \to 0$ *be an exact sequence. For any A-module N, it induces natural exact sequences:*

$$M' \otimes_A N \xrightarrow{g \otimes N} M \otimes_A N \xrightarrow{f \otimes N} M'' \otimes_A N \to 0,$$

$$0 \to \operatorname{Hom}_A(M'', N) \xrightarrow{\operatorname{Hom}_A(g, N)} \operatorname{Hom}_A(M, N) \xrightarrow{\operatorname{Hom}_A(f, N)} \operatorname{Hom}_A(M', N)$$

$$0 \to \operatorname{Hom}_A(N, M') \xrightarrow{\operatorname{Hom}_A(N, f)} \operatorname{Hom}_A(N, M) \xrightarrow{\operatorname{Hom}_A(N, g)} \operatorname{Hom}_A(N, M').$$

The proof of this proposition is not very exciting. A reader who doubts it should produce his own proof.

Note that if $0 \to M' \xrightarrow{f} M \xrightarrow{g} M'' \to 0$ splits, then the following sequences are exact and split:

$$0 \to M' \otimes_A N \xrightarrow{g \otimes N} M \otimes_A N \xrightarrow{f \otimes N} M'' \otimes_A N \to 0,$$

$$0 \to \mathrm{Hom}_A(M'', N) \xrightarrow{\mathrm{Hom}_A(g,N)} \mathrm{Hom}_A(M, N) \xrightarrow{\mathrm{Hom}_A(f,N)} \mathrm{Hom}_A(M', N) \to 0,$$

$$0 \to \mathrm{Hom}_A(N, M') \xrightarrow{\mathrm{Hom}_A(N,f)} \mathrm{Hom}_A(N, M) \xrightarrow{\mathrm{Hom}_A(N,g)} \mathrm{Hom}_A(N, M') \to 0.$$

Proposition 6.16 can also be stated in the following way:

(i) The covariant functor $N \otimes_A .$ from the category of A-modules to itself is right exact.

(ii) The contravariant functor $\mathrm{Hom}_A(., N)$ is left exact.

(iii) The covariant functor $\mathrm{Hom}_A(N, .)$ is left exact.

Consider an ideal $\mathcal{I}$ of A, the exact sequence $0 \to \mathcal{I} \to A \to A/\mathcal{I} \to 0$ and an A-module M. By applying the functor $M \otimes_A$ to the exact sequence, one gets the following easy but important consequence

Corollary 6.17 $M \otimes_A A/\mathcal{I} \simeq M/\mathcal{I}M$.

Definition 6.18

(i) *An A-module P is flat if for each exact sequence of A-modules $M' \to M \to M''$, the complex $M' \otimes_A P \to M \otimes_A P \to M'' \otimes_A P$ is an exact sequence.*

(ii) *An A-module P is faithfully flat if for each complex of A-modules $M' \to M \to M''$, the complex $M' \otimes_A P \to M \otimes_A P \to M'' \otimes_A P$ is exact if and only if the complex $M' \to M \to M''$ is exact.*

Exercises 6.19 Prove the following statements:

1. a free A-module is faithfully flat;

2. the A-algebra $A[X]$ is a faithfully flat;

3. if B is a flat A-algebra, then $B[X]$ is a flat $A[X]$-algebra;

4. the polynomial ring $A[X_1, \ldots, X_n]$ is faithfully flat over A.

We stress here an important property of faithfully flat A-algebras.

Proposition 6.20 *If B is a faithfully flat A-algebra and $\mathcal{I}$ an ideal of A, then $\mathcal{I}B \cap A = \mathcal{I}$.*

Proof First note that $B \otimes_A A/\mathcal{I} \simeq B/\mathcal{I}B$. Since $(\mathcal{I}B \cap A)B = \mathcal{I}B$, it is clear that the complex $0 \to A/\mathcal{I} \to A/(\mathcal{I}B \cap A)$ induces an exact sequence

$$0 \to B \otimes_A A/\mathcal{I} \to B \otimes_A A/(\mathcal{I}B \cap A).$$

This shows that the natural surjective homomorphism $A/\mathcal{I} \to A/(\mathcal{I}B \cap A)$ is an isomorphism.

6.4 Dualizing module on an artinian ring

Definition 6.21 *Let A be an artinian ring. A finitely generated A-module D is dualizing if the natural evaluation application*

$$e_{D,M} : M \to \mathrm{Hom}_A(\mathrm{Hom}_A(M, D), D), \quad e_{D,M}(x)(f) = f(x)$$

is an isomorphism for all finitely generated A-modules M.

Example 6.22 Let k be a field. The dualizing k-modules are the k-vector spaces of rank 1.

Definition 6.23 *If an artinian ring A is a dualizing A-module, A is a Gorenstein artinian ring.*

Note that an artinian ring A is Gorenstein if and only if every finitely generated A-module is reflexive.

Theorem 6.24 *Let A be an artinian ring and D a finitely generated A-module. The following conditions are equivalent:*

(i) *the A-module D is dualizing;*

(ii) *the A-module D is faithful and for all maximal ideals $\mathcal{M}$ of A one has $A/\mathcal{M} \simeq \mathrm{Hom}_A(A/\mathcal{M}, D)$;*

(iii) *the A-module D satisfies $l_A(D) = l_A(A)$ and for all maximal ideals $\mathcal{M}$ of A one has $A/\mathcal{M} \simeq \mathrm{Hom}_A(A/\mathcal{M}, D)$;*

(iv) *for all finitely generated A-modules M, one has*

$$l_A(\mathrm{Hom}_A(M, D)) = l_A(M);$$

(v) *for all injective homomorphisms $N \hookrightarrow M$ of finitely generated A-modules, the natural homomorphism $\mathrm{Hom}_A(M, D) \to \mathrm{Hom}_A(N, D)$ is surjective and for all maximal ideals $\mathcal{M}$ of A, one has $A/\mathcal{M} \simeq \mathrm{Hom}_A(A/\mathcal{M}, D)$.*

Proof

(i) $\Rightarrow$ (ii). Assume D is a dualizing A-module. The isomorphism

$$e_{D,A} : A \simeq \mathrm{Hom}_A(\mathrm{Hom}_A(A, D), D)$$

shows that D is faithful.

Let $\mathcal{M}$ be a maximal ideal of A. If $a \in \mathcal{M}$ and $f \in \mathrm{Hom}_A(A/\mathcal{M}, D)$, we have

$$af(x) = f(ax) = f(0) = 0.$$

Hence $\mathrm{Hom}_A(A/\mathcal{M}, D)$ is an $A/\mathcal{M}$-vector space. If it has rank r, consider an isomorphism $\mathrm{Hom}_A(A/\mathcal{M}, D) \simeq r(A/\mathcal{M})$. It induces isomorphisms

$$\mathrm{Hom}_A(\mathrm{Hom}_A(A/\mathcal{M}, D), D) \simeq \mathrm{Hom}_A(r(A/\mathcal{M}), D)$$

$$\simeq r(\mathrm{Hom}_A(A/\mathcal{M}, D)) \simeq r^2(A/\mathcal{M}).$$

Since D is dualizing, $A/\mathcal{M} \simeq \mathrm{Hom}_A(\mathrm{Hom}_A(A/\mathcal{M}, D), D)$ and $r^2 = 1$. $\square$

Before we go on, let us prove, by induction on $l(M)$, the following assertion.

(*) *If $A/\mathcal{M} \simeq \mathrm{Hom}_A(A/\mathcal{M}, D)$ for all maximal ideals $\mathcal{M}$ of A, then for all finitely generated A-modules M, one has $l_A(\mathrm{Hom}_A(M, D)) \leq l_A(M)$.*

If $l_A(M) = 1$, there is a maximal ideal $\mathcal{M}$ such that $M \simeq A/\mathcal{M}$. Hence

$$l_A(\mathrm{Hom}_A(M, D)) = l_A(\mathrm{Hom}_A(A/\mathcal{M}, D)) = l_A(A/\mathcal{M}) = 1.$$

If $l_A(M) > 1$, Let $M' \subset M$ be a strict submodule. We have $l_A(M') < l_A(M)$ and $l_A(M/M') < l_A(M)$.

The exact sequence

$$0 \to \mathrm{Hom}_A(M/M', D) \to \mathrm{Hom}_A(M, D) \to \mathrm{Hom}_A(M', D)$$

shows

$$l_A(\mathrm{Hom}_A(M, D)) \leq l_A(\mathrm{Hom}_A(M', D)) + l_A(\mathrm{Hom}_A(M/M', D)),$$

from which we deduce, by the induction hypothesis,

$$l_A(\mathrm{Hom}_A(M, D)) \leq l_A(M') + l_A(M/M') = l_A(M).$$

Our assertion being proved, we can go back to the proof of Theorem 6.24.

(ii) $\Rightarrow$ (iii). Using (*) twice, we find

$$l_A(A) \geq l_A(\mathrm{Hom}_A(A, D)) = l_A(D) \geq l_A(\mathrm{Hom}_A(D, D)).$$

Since $((0) : D) = (0)$, the natural homomorphism $A \to \mathrm{Hom}_A(D, D)$ is injective. Consequently $l_A(A) \leq l_A(\mathrm{Hom}_A(D, D))$ and

$$l_A(A) = l_A(\mathrm{Hom}_A(A, D)) = l_A(D) = l_A(\mathrm{Hom}_A(D, D)).$$

(iii) $\Rightarrow$ (iv). By (*), $l_A(\mathrm{Hom}_A(M, D)) \leq l_A(M)$ for all finitely generated A-modules M. If M is finitely generated, there exists an integer n and an exact sequence

$$0 \to K \to nA \to M \to 0.$$

This induces an exact sequence

$$0 \to \mathrm{Hom}_A(M, D) \to \mathrm{Hom}_A(nA, D) \to \mathrm{Hom}_A(K, D)$$

which shows

$$l_A(\mathrm{Hom}_A(nA, D)) \leq l_A(\mathrm{Hom}_A(M, D)) + l_A(\mathrm{Hom}_A(K, D)).$$

Since $\mathrm{Hom}_A(nA, D) = nD$, we have

$$l_A(\mathrm{Hom}_A(nA, D)) = nl_A(D) = nl_A(A) = l_A(M) + l_A(K),$$

hence

$$l_A(M) + l_A(K) \leq l_A(\mathrm{Hom}_A(M, D)) + l_A(\mathrm{Hom}_A(K, D)) \leq l_A(M) + l_A(K),$$

which shows

$$l_A(\mathrm{Hom}_A(M, D)) = l_A(M).$$

(iv) $\Rightarrow$ (v). The exact sequence

$$0 \to N \to M \to M/N \to 0,$$

induces an exact sequence

$$0 \to \mathrm{Hom}_A(N, D) \to \mathrm{Hom}_A(M, D) \to \mathrm{Hom}_A(M/N, D) \to T \to 0,$$

where T is by definition the cokernel of the preceding arrow. Note that

$$\begin{aligned} l_A(\mathrm{Hom}_A(M, D)) &= l_A(M) = l_A(N) + l_A(M/N) \\ &= l_A(\mathrm{Hom}_A(N, D)) + l_A(\mathrm{Hom}_A(M/N, D)). \end{aligned}$$

This shows $l_A(T) = 0$, hence $T = (0)$ and we are done.

(v) $\Rightarrow$ (i). To begin with, note that if the evaluation homomorphism $e_{D,M} = 0$, then $f(x) = 0$ for all $f \in \operatorname{Hom}_A(M, D)$ and all $x \in M$. This shows $\operatorname{Hom}_A(M, D) = (0)$. Let $\mathcal{M}$ be a maximal ideal. Since $\operatorname{Hom}_A(A/\mathcal{M}, D) \neq 0$, then

$$e_{D,A/\mathcal{M}} : A/\mathcal{M} \to \operatorname{Hom}_A(\operatorname{Hom}_A(A/\mathcal{M}, D), D)$$

is different from zero. Note next that

$$A/\mathcal{M} \simeq \operatorname{Hom}_A(A/\mathcal{M}, D) \simeq \operatorname{Hom}_A(\operatorname{Hom}_A(A/\mathcal{M}, D), D).$$

But a non-zero homomorphism from $A/\mathcal{M}$ to $A/\mathcal{M}$ is obviously an isomormorhism. We have proved that $e_{D,A/\mathcal{M}}$ is an isomorphism. Now, if $l_A(M) = 1$, there exists a maximal ideal $\mathcal{M}$ such that $M \simeq A/\mathcal{M}$. This shows that $e_{D,M}$ is an isomorphism.

We can now prove, by induction on $l_A(M)$, that the evaluation homomorphism

$$e_{D,M} : M \to \operatorname{Hom}_A(\operatorname{Hom}_A(M, D), D)$$

is an isomorphism for all finitely generated modules M.

Assume $l_A(M) > 1$. Let $M' \subset M$ be a strict submodule. We have $l_A(M') < l_A(M)$ and $l_A(M/M') < l_A(M)$. Consider the following commutative diagram (where we write $N_D^{\smile\smile}$ for $\operatorname{Hom}_A(\operatorname{Hom}_A(N, D), D)$) :

$$
\begin{array}{ccccccccc}
0 & \to & M' & \to & M & \to & M/M' & \to & 0 \\
& & \downarrow & & \downarrow & & \downarrow & & \\
0 & \to & M_D'^{\smile\smile} & \to & M_D^{\smile\smile} & \to & (M/M')_D^{\smile\smile} & \to & 0.
\end{array}
$$

The first line is exact by hypothesis and the second line is exact by applying twice hypothesis (v). By the induction hypothesis, the vertical arrows $e_{D,M'}$ and $e_{D,M/M'}$ are isomorphisms, hence so is $e_{D,M}$ by the snake lemma. $\qquad\square$

Theorem 6.25 *Let A be an artinian ring. If D and D' are dualizing A-modules, there exists an isomorphism $D \simeq D'$.*

Proof Since $l_A(D) = l_A(D')$, it is enough to find an injective homomorphism $D \to D'$.

Let $\mathcal{M}_1, ..., \mathcal{M}_n$ be the maximal ideals of A. The natural isomorphisms

$$\operatorname{Hom}_A(A/\mathcal{M}_i, D) \simeq (0_D : \mathcal{M}_i), \quad f \to f(1),$$

induce isomorphisms

$$(0_D : \mathcal{M}_i) \simeq A/\mathcal{M}_i.$$

We claim that $(0_D : \mathcal{M}_i) \cap (0_D : \bigcap_{j \neq i} \mathcal{M}_j) = (0)$. Indeed, if $x \in (0_D : \mathcal{M}_i) \cap (0_D : \bigcap_{j \neq i} \mathcal{M}_j)$, we have $x\mathcal{M}_i = x(\bigcap_{j \neq i} \mathcal{M}_j) = (0)$. Since $\mathcal{M}_i$ and $\bigcap_{j \neq i} \mathcal{M}_j$ are comaximal, this shows $xA = 0$, hence $x = 0$. As a consequence,

$$\sum_1^n (0_D : \mathcal{M}_i) \simeq \bigoplus_1^n (0_D : \mathcal{M}_i) \simeq \bigoplus_1^n A/\mathcal{M}_i$$

and

$$\sum_1^n (0_{D'} : \mathcal{M}_i) = \bigoplus_1^n (0_{D'} : \mathcal{M}_i) \simeq \bigoplus_1^n A/\mathcal{M}_i.$$

Next note that if $K \subset D$, then

$$K \cap \left(\sum_1^n (0_D : \mathcal{M}_i)\right) = (0) \Rightarrow K = (0).$$

Indeed, if $K \neq (0)$, there exists i such that $\mathcal{M}_i \in \text{Ass}(K)$. This implies $(0_K : \mathcal{M}_i) \neq (0)$ and since $(0_K : \mathcal{M}_i) \subset (0_D : \mathcal{M}_i)$, we are done.

The isomorphism $\sum_1^n (0_D : \mathcal{M}_i) \simeq \sum_1^n (0_{D'} : \mathcal{M}_i)$ induces an injective homomorphism

$$f : \sum_1^n (0_D : \mathcal{M}_i) \to D'.$$

By Theorem 6.24 (v), the inclusion $\sum_1^n (0_D : \mathcal{M}_i) \subset D$ induces a surjective homomorphism

$$\text{Hom}_A(D.D') \to \text{Hom}_A\left(\sum_1^n (0_D : \mathcal{M}_i), D'\right).$$

Hence there exists a homomorphism $g : D \to D'$ whose restriction to $\sum_1^n (0_D : \mathcal{M}_i)$ is f. This shows

$$\ker g \cap \sum_1^n (0_D : \mathcal{M}_i) = \ker f = (0),$$

hence $\ker g = (0)$, and the theorem is proved. $\qquad\qquad\qquad\square$

Theorem 6.26 *Let A be an artinian ring and B an A-algebra finitely generated as an A-module. If D is a dualizing A-module, then $\text{Hom}_A(B, D)$ is a dualizing B-module.*

Proof Recall first that B is artinian by Corollary 5.22 and note next that $\text{Hom}_A(B, D)$ has a unique natural structure of a B-module.

We use Theorem 6.24 (ii). To prove that $\text{Hom}_A(B, D)$ is a faithful B-module, consider a non-zero element $b \in B$. Since the evaluation homomorphism $B \simeq \text{Hom}_A(\text{Hom}_A(B, D), D)$ is injective, there exists $f \in \text{Hom}_A(B, D)$ such that $f(b) \neq 0$. In other words $bf \neq 0$.

Next, we let $\mathcal{N}$ be a maximal ideal of B. What we want to prove is that $\operatorname{Hom}_B(B/\mathcal{N}, \operatorname{Hom}_A(B, D))$ is a $B/\mathcal{N}$-vector space of rank 1.

Put $\mathcal{M} = \mathcal{N} \cap A$. Note that $B/\mathcal{N}$ is a finitely generated A-module annihilated by $\mathcal{M}$, hence a finite rank $A/\mathcal{M}$-vector space.

The natural homomorphism (of A-modules) $\operatorname{Hom}_A(B, D) \to D$, $f \to f(1)$ induces a homomorphism of A-modules

$$\operatorname{Hom}_B(B/\mathcal{N}, \operatorname{Hom}_A(B, D)) \to \operatorname{Hom}_A(B/\mathcal{N}, D).$$

It is straightforward to check that this is an isomorphism of A-modules, hence of $A/\mathcal{M}$-vector spaces. Using the fact that D is a dualizing A-module, we get as a consequence

$$l_A(\operatorname{Hom}_B(B/\mathcal{N}, \operatorname{Hom}_A(B, D))) = l_A(\operatorname{Hom}_A(B/\mathcal{N}, D)) = l_A(B/\mathcal{N}).$$

Note that if P is a finitely generated A-module annihilated by $\mathcal{M}$, hence an $A/\mathcal{M}$-vector space, we have $l_A(P) = \operatorname{rk}_{A/\mathcal{M}}(P)$. Therefore we have proved

$$\operatorname{rk}_{A/\mathcal{M}}(\operatorname{Hom}_B(B/\mathcal{N}, \operatorname{Hom}_A(B, D))) = \operatorname{rk}_{A/\mathcal{M}}(B/\mathcal{N}).$$

By Proposition 5.24, this implies

$$\operatorname{rk}_{B/\mathcal{N}}(\operatorname{Hom}_B(B/\mathcal{N}, \operatorname{Hom}_A(B, D))) = \operatorname{rk}_{B/\mathcal{N}}(B/\mathcal{N}) = 1,$$

and the theorem is proved. $\qquad\square$

6.5 Gorenstein artinian rings

We recall that an artinian ring A such that A is a dualizing A-module is called Gorenstein. We recall also, and this is an important characterization, that an artinian ring A is Gorenstein if and only if all finitely generated A-modules are reflexive.

Theorem 6.27 *Let A be an artinian ring and $(0) = \bigcap_1^n \mathcal{Q}_i$ the minimal primary decomposition. Then A is Gorenstein if only if $\mathcal{Q}_i$ is irreducible for all i.*

Proof Put $\mathcal{M}_i = \sqrt{\mathcal{Q}_i}$. This is a maximal ideal and all maximal ideals of A are thus obtained. Since A is a faithful A-module, by Theorem 6.24 it is enough to show that $\mathcal{Q}_i$ is irreducible if and only if $\operatorname{Hom}_A(A/\mathcal{M}_i, A) \simeq A/\mathcal{M}_i$.

By Theorem 1.62, we have $A \simeq \prod_1^n A/\mathcal{Q}_i$. We claim that for $i \neq j$,

$$\operatorname{Hom}_A(A/\mathcal{M}_i, A/\mathcal{Q}_j) = (0).$$

Indeed the ideals $\mathcal{M}_i$ and $\mathcal{Q}_j$ are comaximal and are both contained in the annihilator of this module. This shows

$$\mathrm{Hom}_A(A/\mathcal{M}_i, A) \simeq \mathrm{Hom}_A(A/\mathcal{M}_i, A/\mathcal{Q}_i) \simeq (\mathcal{Q}_i : \mathcal{M}_i)/\mathcal{Q}_i.$$

Consider the $A/\mathcal{M}_i$-vector space $(\mathcal{Q}_i : \mathcal{M}_i)/\mathcal{Q}_i$. If $\mathcal{Q}_i$ is irreducible, then (0) is irreducible in the ring $A/\mathcal{Q}_i$. Since $(\mathcal{Q}_i : \mathcal{M}_i)/\mathcal{Q}_i$ is an ideal of this ring, (0) is irreducible in this vector space which is therefore of rank 1. Hence we have proved

$$\mathrm{Hom}_A(A/\mathcal{M}_i, A) \simeq (\mathcal{Q}_i : \mathcal{M}_i)/\mathcal{Q}_i \simeq A/\mathcal{M}_i.$$

Conversely, assume $(\mathcal{Q}_i : \mathcal{M}_i)/\mathcal{Q}_i \simeq A/\mathcal{M}_i$. To prove that $\mathcal{Q}_i$ is irreducible, we show that if $\mathcal{I}$ is an ideal strictly containing $\mathcal{Q}_i$, then $(\mathcal{Q}_i : \mathcal{M}_i) \subset \mathcal{I}$.

We have $\mathrm{Ass}(\mathcal{I}/\mathcal{Q}_i) \subset \mathrm{Ass}(A/\mathcal{Q}_i) = \{\mathcal{M}_i\}$. This proves

$$(\mathcal{I}/\mathcal{Q}_i) \cap ((\mathcal{Q}_i : \mathcal{M}_i)/\mathcal{Q}_i) \neq (0).$$

Since $(\mathcal{Q}_i : \mathcal{M}_i)/\mathcal{Q}_i$ is an $A/\mathcal{M}_i$-vector space of rank 1, we must have

$$(\mathcal{Q}_i : \mathcal{M}_i)/\mathcal{Q}_i \subset \mathcal{I}/\mathcal{Q}_i.$$

$\square$

Corollary 6.28 *If R is a principal ideal ring and $x \neq 0$ a non-invertible element of R, then R/xR is a Gorenstein artinian ring.*

Proof Let $x = p_1^{n_1}...p_r^{n_r}$ be a decomposition of x in prime factors. The prime ideals of R/xR are $p_i R/xR$. They are all maximal, hence R/xR is artinian.

Furthermore

$$(0) = \bigcap_1^r (p_i^{n_i} R/xR)$$

is a minimal primary decomposition of (0) in R/xR. Showing that $p_i^{n_i} R$ is an irreducible ideal of R is an excellent exercise. Do it. $\square$

Exercise 6.29 Let $P \in \mathbb{C}[X]$ be a non-zero polynomial of degree $n > 0$. The ring $A = \mathbb{C}[X]/(P)$ is a $\mathbb{C}$-vector space of rank n. The elements $\mathrm{cl}(X^j) \in A$, with $0 \leq j \leq n-1$, form a basis of this $\mathbb{C}$-vector space.

By Corollary 6.28, A is a Gorenstein ring, hence A is a dualizing A-module. By Theorem 6.26, the A-module $\mathrm{Hom}_{\mathbb{C}}(A, \mathbb{C})$ is also dualizing, hence is isomorphic to A by Theorem 6.25.

Show that the element $\tau \in \mathrm{Hom}_{\mathbb{C}}(A, \mathbb{C})$, defined by $\tau(\mathrm{cl}(X^j)) = \delta_{j,n-1}$, is a basis of this rank-one free A-module.

Theorem 6.30 *Let A be a Gorenstein artinian ring and $\mathcal{I}$ an ideal of A. The following conditions are equivalent:*

(i) *the quotient ring $A/\mathcal{I}$ is Gorenstein;*

(ii) *the ideal $(0) : \mathcal{I}$ is principal;*

(iii) *there exists an element $a \in A$ such that $(0 : a) = \mathcal{I}$.*

Proof We recall first that for any ideal $\mathcal{J}$ of a ring R, there is a natural isomorphism $\mathrm{Hom}_R(R/\mathcal{J}, R) \simeq ((0) : \mathcal{J})$.

We know, from Theorem 6.26, that $\mathrm{Hom}_A(A/\mathcal{I}, A)$ is a dualizing module for $A/\mathcal{I}$. Hence if $A/\mathcal{I}$ is Gorenstein, then $((0) : \mathcal{I}) \simeq \mathrm{Hom}_A(A/\mathcal{I}, A) \simeq A/\mathcal{I}$. This shows that $((0) : \mathcal{I})$ is a principal ideal and consequently that (i) implies (ii).

Assume now $\mathrm{Hom}_A(A/\mathcal{I}, A) \simeq ((0) : \mathcal{I}) = aA \simeq A/(0 : a)$. Since A is a dualizing A-module, we have $l_A(\mathrm{Hom}_A(M, A)) = l_A(M)$, for all finitely generated A-modules M, by Theorem 6.24 (iv). This implies

$$l_A(A/\mathcal{I}) = l_A(\mathrm{Hom}_A(A/\mathcal{I}, A)) = l_A(A/(0 : a)).$$

But obviously, $\mathcal{I} \subset (0 : a)$ and $A/(0 : a)$ is a quotient ring of $A/\mathcal{I}$. Since they have the same length, this shows that $A/(0 : a) = A/\mathcal{I}$, hence that $(0 : a) = \mathcal{I}$ and (ii) implies (iii).

Assume next $(0 : a) = \mathcal{I}$. This implies $aA \simeq A/\mathcal{I}$, hence $l_A(aA) = l_A(A/\mathcal{I})$. On the other hand, we have

$$aA \subset ((0) : \mathcal{I}) \simeq \mathrm{Hom}_A(A/\mathcal{I}, A),$$

hence $l_A(aA) \leq l_A(\mathrm{Hom}_A(A/\mathcal{I}, A))$. But, by Theorem 6.24 (iv), we know that $l_A(A/\mathcal{I}) = l_A(\mathrm{Hom}_A(A/\mathcal{I}, A))$. This proves $l_A(aA) = l_A(\mathrm{Hom}_A(A/\mathcal{I}, A))$ and therefore

$$aA = ((0) : \mathcal{I}) \simeq \mathrm{Hom}_A(A/\mathcal{I}, A).$$

Finally, we get

$$A/\mathcal{I} \simeq aA = ((0) : \mathcal{I}) \simeq \mathrm{Hom}_A(A/\mathcal{I}, A).$$

This shows that $A/\mathcal{I}$ is a dualizing $A/\mathcal{I}$-module. Consequently $A/\mathcal{I}$ is Gorenstein and (iii) implies (i). $\qquad \square$

6.6 Exercises

1. Let A be a ring and $a, b \in A$. Show that the complex

$$0 \to A \xrightarrow{\binom{b}{-a}} 2A \xrightarrow{(a \ b)} A$$

is exact if and only if $a \in A$ and $\mathrm{cl}(b) \in A/aA$ are not zero divisors (or if $b \in A$ and $\mathrm{cl}(a) \in A/bA$ are not zero divisors).

2. Let A be a ring, M an A-module and $a \in A$ an element regular in M, i.e. such that $x \in M$ and $ax = 0 \Rightarrow x = 0$. Show that if M/aM has finite length, then
$$l(M/a^r M) = rl(M/aM).$$

3. Consider the ring $A = \mathbb{C}[X, Y, Z]$ and show that the following complex is exact:
$$0 \to A \xrightarrow{\begin{pmatrix} X \\ Y \\ Z \end{pmatrix}} 3A \xrightarrow{\begin{pmatrix} 0 & -Z & Y \\ Z & 0 & -X \\ -Y & X & 0 \end{pmatrix}} 3A \xrightarrow{\begin{pmatrix} X & Y & Z \end{pmatrix}} A$$

4. Let R be a domain and $a, b \in R$ non-zero elements. Show that the following complex is exact:
$$\dots \to R/abR \xrightarrow{a} R/abR \xrightarrow{b} R/abR \xrightarrow{a} R/abR \xrightarrow{b} R/abR \to \dots$$

5. Consider the following commutative diagram of modules and homomorphisms:
$$
\begin{array}{ccccccccc}
\dots & \to & M_r & \to & \dots & \to & M_1 & \to & M_0 & \to & 0 \\
 & & \downarrow f_r & & & & \downarrow f_1 & & \downarrow f_0 & & \\
\dots & \to & N_r & \to & \dots & \to & N_1 & \to & N_0 & \to & \dots
\end{array}
$$

Assume that the lines are exact and that f_i is injective for all i. Show that it induces a long exact sequence
$$\dots \to \operatorname{coker} f_r \to \dots \to \operatorname{coker} f_1 \to \operatorname{coker} f_0 \to \dots$$

6. Show that the local ring $\mathbb{C}[X, Y]/(X^n, Y^m)$ is artinian Gorenstein. Hint: show that the ideal (X^n, Y^m) is irreducible.

7. Let A be a Noetherian ring. Assume that $a \in A$ is not a zero divisor and that A/aA is an Artinian Gorenstein ring. Show that $A/a^n A$ is an Artinian Gorenstein ring for all positive integers n.

8. Let A be a UFD. Consider the matrix $M = \begin{pmatrix} a & b & c \\ a' & b' & c' \end{pmatrix}$ with coefficients in A. Put $\Delta_1 = bc' - b'c$, $\Delta_2 = ca' - c'a$, $\Delta_3 = ab' - a'b$ and show

$$M \begin{pmatrix} \Delta_1 \\ \Delta_2 \\ \Delta_3 \end{pmatrix} = 0.$$

Assume that Δ_1, Δ_2 and Δ_3 have no common factor and show that

$$M \begin{pmatrix} d_1 \\ d_2 \\ d_3 \end{pmatrix} = 0 \quad \text{if and only if there exists } d' \in A \text{ such that}$$

$$\begin{pmatrix} d_1 \\ d_2 \\ d_3 \end{pmatrix} = d' \begin{pmatrix} \Delta_1 \\ \Delta_2 \\ \Delta_3 \end{pmatrix}.$$

7

Fractions

We are used to representing rational numbers as fractions of integers. We often choose to work with a reduced representation of a fraction. This is possible because $\mathbb{Z}$ is a UFD. We shall see that in a more general setting a fraction does not necessarily have a reduced representation. Defining fraction rings when the ring has zero divisors and then fraction modules takes a bit of work. This is a first step towards the theory of sheaves, so central in geometry. We proceed further in this direction by introducing the support of a module. From section 4 on, we study the relations between the ideals of a ring and the ideals of its fraction rings, with a special interest in prime ideals. Using these new techniques, we prove, eventually, that a polynomial ring over a UFD is a UFD. In the Noetherian case we understand how fractions and primary decomposition interact.

7.1   Rings of fractions

Consider in $\mathbb{Z} \times (\mathbb{Z} \setminus \{0\})$ the equivalence relation

$$(a, s) \sim (b, t) \quad \text{if} \quad at - bs = 0.$$

Denote by a/s the class of (a, s). It is clear that in the set of equivalence classes the operations

$$a/s + b/t = (at + bs)/st \quad \text{and} \quad (a/s)(b/t) = ab/st$$

are well defined. They give to this set the structure of a field, with $0 = 0/s$ and $1 = 1/1$.

This field is obviously $\mathbb{Q}$. The map

$$\mathrm{i} : \mathbb{Z} \to \mathbb{Q}, \quad \mathrm{i}(a) = a/1,$$

is, as obviously, the natural inclusion of $\mathbb{Z}$ as a subring of $\mathbb{Q}$.

Definition 7.1 *A part S of a ring A is called multiplicatively closed if*

$$1 \in S \quad \text{and} \quad s, s' \in S \implies ss' \in S.$$

Note that $\mathbb{Z} \setminus \{0\}$ is a multiplicatively closed part of $\mathbb{Z}$.

Definition 7.2 *Let A be a ring and S be a multiplicatively closed part of it. We denote by $S^{-1}A$ the quotient of $A \times S$ by the equivalence relation $(a, s) \sim (b, t)$ if there exists $r \in S$ such that $r(at - bs) = 0$ and by $a/s \in S^{-1}A$ the class of (a, s).*

The proof of the following theorem is straightforward.

Theorem 7.3 (i) *The operations*

$$a/s + b/t = (at + bs)/st \quad \text{and} \quad (a/s)(b/t) = ab/st$$

are well defined in $S^{-1}A$.

(ii) *Equipped with these operations, $S^{-1}A$ is a commutative ring.*

(iii) *The map $\mathrm{i} : A \to S^{-1}A$, $\mathrm{i}(a) = a/1$, is a ring homomorphism.*

(iv) *The kernel of this homomorphism is the ideal $\bigcup_{s \in S}((0) : s)$.*

We make some obvious comments before going on.

- The set $\bigcup_{s \in S}((0) : s)$ is an ideal because S is multiplicatively closed.

- We have $S^{-1}A = 0$ if and only if $1 \in \ker \mathrm{i}$, in other words if and only if $0 \in S$.

- The canonical ring homomorphism $\mathrm{i} : A \to S^{-1}A$ is an isomorphism if and only if all $s \in S$ are units.

- The canonical ring homomorphism $\mathrm{i} : A \to S^{-1}A$ is injective if and only if S does not contain any zero divisor. When this is the case, we will often consider A as a subring of $S^{-1}A$.

- If $T \subset S$ is also a multiplicatively closed part, the homomorphism $A \to S^{-1}A$ factorizes through the homomorphism $A \to T^{-1}A$ and a natural homomorphism $T^{-1}A \to S^{-1}A$.

Examples 7.4 Assume that A is a domain.

1. The part $S = A \setminus \{0\}$ of all non-zero elements is multiplicatively closed.

2. The ring $S^{-1}A$ is a field, the fraction field of A, often denoted by $K(A)$.

3. The canonical ring homomorphism $\mathrm{i} : A \to K(A)$ is injective.

4. For any multiplicatively closed part T, with $0 \notin T$, the ring $T^{-1}A$ is naturally a subring of the fraction field $K(A)$.

Exercises 7.5

1. If $s \in A$ is not nilpotent and $S = \{s^n\}_{n \geq 0}$, one often writes A_s for $S^{-1}A$. Show that the A-algebra homomorphism $A[X] \to A_s$, $P(X) \to P(1/s)$ induces an isomorphism $A[X]/(sX - 1) \simeq A_s$.

2. If $\mathcal{P}$ is a prime ideal of A, then $S = A \setminus \mathcal{P}$ is multiplicatively closed. Show that all multiplicatively closed parts of A are not thus obtained (hint: $A \setminus S$ is not necessarily an ideal).

Definition 7.6 *If $\mathcal{P}$ is a prime ideal of A and $S = A \setminus \mathcal{P}$, we put*

$$A_\mathcal{P} = S^{-1}A.$$

Note that $A_\mathcal{P} \neq (0)$ for all prime ideals $\mathcal{P}$ of A.

Example 7.7 Let $\mathcal{M} = (X_1 - a_1, ..., X_n - a_n) \subset \mathbb{C}[X_1, ..., X_n]$ be the maximal ideal whose elements are the polynomials $P \in \mathbb{C}[X_1, ..., X_n]$ such that $P(a_1, ..., a_n) = 0$. Then $A_\mathcal{M}$ is the ring of rational functions defined at the point $(a_1, ..., a_n) \in \mathbb{C}^n$.

Theorem 7.8 *Let S be a multiplicatively closed part of A and $\mathrm{i} : A \to S^{-1}A$ the canonical ring homomorphism.*

(i) *If $s \in S$, the element $\mathrm{i}(s) \in S^{-1}A$ is invertible.*

(ii) *If $f : A \to B$ is a ring homomorphism such that $f(s)$ is invertible for all $s \in S$, there exists a unique ring homomorphism $\overline{f} : S^{-1}A \to B$ such that $f = \overline{f} \circ \mathrm{i}$.*

Proof That $\mathrm{i}(s)1/s = 1/1$ is obvious. Define $\overline{f}(a/s) = f(a)s^{-1}$, then (ii) follows. $\qquad \square$

Example 7.9 Let A be a domain and $f : A \to K$ an injective homomorphism with values in a field. There is a natural factorization of f through the fractions field $K(A)$ of A.

Indeed, since f is injective, $f(s)$ is invertible for all $s \neq 0$.

Proposition 7.10 *Let A be a domain. If $\mathrm{Spec}_\mathrm{m}A$ is the set of all maximal ideals of A, then*

$$A = \bigcap_{\mathcal{M} \in \mathrm{Spec}_\mathrm{m}A} A_\mathcal{M}.$$

Proof Note first that if $K(A)$ is the fraction field of A, then $A \subset A_\mathcal{M} \subset K(A)$. Consider next $x/y \in \bigcap_{\mathcal{M} \in \mathrm{Spec}_\mathrm{m}A} A_\mathcal{M}$. Clearly $x \in yA_\mathcal{M}$ for all maximal ideals $\mathcal{M}$. This shows $(yA : x) \not\subset \mathcal{M}$ for all maximal ideals $\mathcal{M}$, hence $(yA : x) = A$ and $x/y \in A$. $\qquad \square$

7.2   Fraction modules

Definition 7.11 *Let S be a multiplicatively closed part of the ring A and M an A-module.*

We denote by $S^{-1}M$ the quotient of $M \times S$ by the equivalence relation $(x, s) \sim (y, t)$ if there exists $r \in S$ such that $r(xt - ys) = 0$ and by $x/s \in S^{-1}M$ the class of (x, s).

The proof of the following theorem is straightforward.

Theorem 7.12 (i) *The operations $x/s + y/t = (tx + sy)/st$ (for $x, y \in M$ and $s, t \in S$) and $(a/s)(x/t) = ab/st$ (for $x \in M$, $a \in A$ and $s, t \in S$) are well defined.*

(ii) *Equipped with these operations $S^{-1}M$ is an $S^{-1}A$-module (hence an A-module as well).*

(iii) *The map $1_M : M \to S^{-1}M$, $1_M(x) = x/1$, is a homomorphism of A-modules.*

(iv) *The kernel of this homomorphism is $\bigcup_{s \in S}(0_M : s)$.*

The map 1_M is often called the localization map.

Examples 7.13

1. If $s \in A$ and $S = \{s^n\}_{n \geq 0}$, we denote $S^{-1}M$ by M_s.

2. If $S = A \setminus \mathcal{P}$ where $\mathcal{P}$ is a prime ideal of A, we denote $S^{-1}M$ by $M_{\mathcal{P}}$.

To an A-module M, we have associated the $S^{-1}A$-module $S^{-1}M$. Next, to a homomorphism of A-modules we associate a homomorphism of $S^{-1}A$-modules. We define thus a "covariant functor from the category of A-modules to the category of $S^{-1}A$-modules". We also prove that this functor is "exact".

Proposition 7.14

(i) *If $\phi : M \to N$ is a homomorphism of A-modules, the map*

$$S^{-1}\phi : S^{-1}M \to S^{-1}N, \quad S^{-1}\phi(x/s) = \phi(x)/s$$

is well defined and is a homomorphism of $S^{-1}A$-modules.

(ii) *If ϕ and ψ are composable homomorphisms of A-modules, then*

$$S^{-1}(\phi \circ \psi) = S^{-1}\phi \circ S^{-1}\psi.$$

(iii) *If* $M \overset{\psi}{\to} N \overset{\phi}{\to} P$ *is an exact sequence of A-modules, then*

$$S^{-1}M \overset{S^{-1}\psi}{\to} S^{-1}N \overset{S^{-1}\phi}{\to} S^{-1}P$$

is an exact sequence of $S^{-1}A$-modules.

Proof (i) and (ii) are obvious. We show (iii). If $x/s \in \ker S^{-1}\phi$, then $\phi(x)/s = 0$. Hence there exists $t \in S$ such that $t\phi(x) = 0 = \phi(tx)$. Let $y \in M$ be such that $tx = \psi(y)$. We have $x/s = \psi(y/st)$. $\qquad\qquad\Box$

The proofs of the next six results are straightforward and left to the reader.

Proposition 7.15 *If S and T are multiplicatively closed parts of A, the part U of all st, with $s \in S$ and $t \in T$, is multiplicatively closed. If M is an A-module, there are natural isomorphisms of $U^{-1}A$-modules*

$$U^{-1}M \simeq S^{-1}(T^{-1}M) \simeq T^{-1}(S^{-1}M).$$

Corollary 7.16 *If $\mathcal{P}$ is a prime ideal, then $A_{\mathcal{P}}/\mathcal{P}A_{\mathcal{P}}$ is a field and $M_{\mathcal{P}}/\mathcal{P}M_{\mathcal{P}}$ a vector space on this field.*

Proposition 7.17 *Let S be a multiplicatively closed part of A and M an A-module.*

(i) *If N is a submodule of M, then $S^{-1}N$ has the natural structure of an $S^{-1}A$-submodule of $S^{-1}M$ and there is a natural isomorphism $S^{-1}M/S^{-1}N \simeq S^{-1}(M/N)$.*

(ii) *If N' is another submodule of M, then*

$$S^{-1}(N + N') = S^{-1}N + S^{-1}N' \quad and \quad S^{-1}(N \cap N') = S^{-1}N \cap S^{-1}N'.$$

(iii) *If F is an $S^{-1}A$-submodule of $S^{-1}M$, we denote by $F \cap M$ the submodule of M formed by all $x \in M$ such that $x/1 \in F \subset S^{-1}M$ and we have*

$$F = S^{-1}(F \cap M).$$

(iv) *If N is a submodule of M, then*

$$S^{-1}N \cap M = \{x \in M, \quad \text{there exists } s \in S \text{ such that } sx \in N\}.$$

Yes, we know that the natural map $M \to S^{-1}M$ is not always injective and that our notation $F \cap M$ does not make sense. But this is a truly convenient notation and we should stick to it.

Proposition 7.18 *If $(x_i)_{i \in I}$ generate the A-module M, then $(x_i/1)_{i \in I}$ generate the $S^{-1}A$-module $S^{-1}M$. In particular if M is a finitely generated A-module, then $S^{-1}M$ is a finitely generated $S^{-1}A$-module.*

Corollary 7.19 *If M is Noetherian, then $S^{-1}M$ is Noetherian. In particular a fraction ring of a Noetherian ring is Noetherian.*

Exercise 7.20 Let A be a Noetherian domain and $S = A \setminus (0)$. If M is a finitely generated A-module, then $S^{-1}M$ is a finite rank vector space on the field $S^{-1}A$. Show that the function $\lambda(M) = \mathrm{rk}_{S^{-1}A}(S^{-1}M)$, defined in the category of finitely generated A-modules and with value in $\mathbb{Z}$, is additive.

Proposition 7.21 *If M and N are A-modules and S a multiplicatively closed part of A, there are a natural isomorphisms of $S^{-1}A$-modules:*

$$S^{-1}A \otimes_A M \simeq S^{-1}M \quad and \quad S^{-1}M \otimes_{S^{-1}A} S^{-1}N \simeq S^{-1}(M \otimes_A N).$$

The geometric consequences of the following proposition and its corollary are important. We will use them in proving the semi-continuity theorem 7.33.

Proposition 7.22 *Let M be a finitely generated A-module and S a multiplicatively closed part of A. The following conditions are equivalent:*

(i) $S^{-1}M = (0)$;

(ii) *there exists $s \in S$ such that $sM = (0)$;*

(iii) *there exists $s \in S$ such that $M_s = (0)$.*

Proof Assume $S^{-1}M = (0)$. Let $x_1, ..., x_n \in M$ be generators of M. Since $x_i/1 = 0 \in S^{-1}M$, there exists $s_i \in S$ such that $s_i x_i = 0$. If $s = s_1...s_n$, we have $sx_i = 0$ for all i, hence $sM = (0)$. Now $sM = (0)$ obviously implies $M_s = (0)$; and just as obviously, $M_s = (0)$, with $s \in S$, implies $S^{-1}M = (0)$.
□

Corollary 7.23 *Let M be a finitely generated A-module and S a multiplicatively closed part of A.*

(i) *If $x_1, ..., x_n \in M$ are elements whose images in $S^{-1}M$ generate $S^{-1}M$ as an $S^{-1}A$-module, there exists $s \in S$ such that the images of $x_1, ..., x_n$ in M_s generate M_s as an A_s-module.*

(ii) *If A is Noetherian and if $S^{-1}M$ is a free $S^{-1}A$-module of rank n, there exists $s \in S$ such that M_s is a free A_s-module of rank n.*

Proof (i) Let C be the cokernel of the homomorphism

$$f : nA \to M, \quad (a_1, ..., a_n) \to \sum a_i x_i.$$

We have assumed $S^{-1}C = (0)$. Since C is finitely generated, there exists, by Proposition 7.22, $s \in S$ such that $C_s = 0$. This shows that f_s is surjective, hence that $x_i/1 \in M_s$ generate M_s.

(ii) Assume $x_1/s_1, ..., x_n/s_n \in S^{-1}M$ form a basis of $S^{-1}M$. One immediately checks that $x_1/1, ..., x_n/1 \in S^{-1}M$ form a basis of $S^{-1}M$. Let K and C be the kernel and the cokernel of the homomorphism f. We have $S^{-1}K = S^{-1}C = (0)$. Since A is Noetherian, the submodule K of nA is finitely generated. Hence there exist t and u in S such that $K_t = (0)$ and $C_u = (0)$. If we put $s = tu$, we have $K_s = C_s = (0)$, hence f_s is an isomorphism and M_s is a free A_s-module of rank n. $\square$

Our last result in this section is important. We will need it several times when studying Weil divisors.

Proposition 7.24 (i) *If M and N are A-modules and S a multiplicatively closed part of A, there is a natural homomorphism of $S^{-1}A$-modules:*

$$S^{-1}\mathrm{Hom}_A(M, N) \to \mathrm{Hom}_{S^{-1}A}(S^{-1}M, S^{-1}N).$$

(ii) *If M is finitely generated, this homomorphism is injective.*

(iii) *If furthermore A is Noetherian, this homomorphism is an isomorphism.*

Proof (i) The homomorphism

$$i : S^{-1}\mathrm{Hom}_A(M, N) \to \mathrm{Hom}_{S^{-1}A}(S^{-1}M, S^{-1}N)$$

is defined by $i(f/s)(x/t) = f(x)/st$ and needs no comment.

(ii) Assume now that M is generated by $x_1, ..., x_n$. If $i(f/s) = 0$, then $i(f/1) = 0$ and $f(x_i)/1 = 0 \in S^{-1}N$ for $i = 1, ..., n$. Consequently, there exists $t \in S$ such that $tf(x_i) = 0$ for $i = 1, ..., n$. This shows $tf(M) = (0)$, hence $f/1 = 0$ and $f/s = 0$.

(iii) Assume now that A is Noetherian. The kernel K of a surjective homomorphism $nA \to M$ is a finitely generated A-module. The exact sequence

$$0 \to \mathrm{Hom}_A(M, N) \to \mathrm{Hom}_A(nA, N) \to \mathrm{Hom}_A(K, N)$$

induces a commutative diagram :

$$
\begin{array}{ccc}
0 & & 0 \\
\downarrow & & \downarrow \\
S^{-1}\mathrm{Hom}_A(M, N) & \to & \mathrm{Hom}_{S^{-1}A}(S^{-1}M, S^{-1}N) \\
\downarrow & & \downarrow \\
S^{-1}\mathrm{Hom}_A(nA, N) & \to & \mathrm{Hom}_{S^{-1}A}(nS^{-1}A, S^{-1}N) \\
\downarrow & & \downarrow \\
S^{-1}\mathrm{Hom}_A(K, N) & \to & \mathrm{Hom}_{S^{-1}A}(S^{-1}K, S^{-1}N).
\end{array}
$$

By (ii), the three horizontal arrows are injective. We claim furthermore that the second, i.e. the homomorphism

$$S^{-1}\mathrm{Hom}_A(nA, N) \to \mathrm{Hom}_{S^{-1}A}(nS^{-1}A, S^{-1}N)$$

is also surjective. Indeed, let $(e_1, ..., e_n)$ be a basis for nA. Consider $f \in \mathrm{Hom}_{S^{-1}A}(nS^{-1}A, S^{-1}N)$ and put $f(e_i/1) = y_i/s_i \in S^{-1}N$. Then if $s = s_1...s_n$, we have $sf(e_i/1) = z_i/1$, with $z_i \in N$. If we define $g \in \mathrm{Hom}_A(nA, N)$ by $g(e_i) = z_i$, we have $f = g/s \in \mathrm{Hom}_{S^{-1}A}(nS^{-1}A, S^{-1}N)$. With this in mind, an easy chase round the diagram proves (iii) in our proposition. $\square$

7.3   Support of a module

We recall that to a ring A, we have associated a topological space $\mathrm{Spec}(A)$. To an A-module M, we associate now a subset $\mathrm{Supp}(M)$ of this topological space. We shall see that when this module is finitely generated, this subset is closed. In this section we study the relations between the module and this subset of $\mathrm{Spec}(A)$.

Definition 7.25 *The support* $\mathrm{Supp}(M) \subset \mathrm{Spec}(A)$ *of an A-module M is the set of prime ideals $\mathcal{P}$ of A such that $M_\mathcal{P} \neq (0)$. We denote by* $\mathrm{Supp}_m(M)$ *the set of all maximal ideals in* $\mathrm{Supp}(M)$

Proposition 7.26 **(i)** $\mathrm{Supp}(A) = \mathrm{Spec}(A)$.

(ii) *If* $0 \to M' \to M \to M'' \to 0$ *is an exact sequence of A-modules, then*

$$\mathrm{Supp}(M) = \mathrm{Supp}(M') \cup \mathrm{Supp}(M'').$$

Proof We have already seen (i). For all prime ideals $\mathcal{P}$ there is an exact sequence

$$0 \to M'_\mathcal{P} \to M_\mathcal{P} \to M''_\mathcal{P} \to 0.$$

This proves (ii). $\square$

Proposition 7.27 *Let M be an A-module. The following conditions are equivalent:*

(i) $M = 0$;

(ii) $\mathrm{Supp}(M) = \emptyset$;

(iii) $\mathrm{Supp}_m(M) = \emptyset$.

Proof It is clearly sufficient to prove that if $M_{\mathcal{M}} = (0)$ for all maximal ideals $\mathcal{M}$ of A, then $M = (0)$.

Let $x \in M$. Since x has a trivial image in $M_{\mathcal{M}}$ there exists $s \notin \mathcal{M}$ such that $sx = 0$. Hence $((0) : x) \not\subset \mathcal{M}$ for all maximal ideals $\mathcal{M}$ of A. This proves $((0) : x) = A$ and $x = 0$. □

Corollary 7.28 *Let $f : M \to N$ be an A-modules homomorphism. The following conditions are equivalent:*

(i) *f is injective (resp. surjective, bijective);*

(ii) *for all prime ideals $\mathcal{P} \in \mathrm{Supp}(M)$ (resp. $\mathcal{P} \in \mathrm{Supp}(N)$, $\mathcal{P} \in \mathrm{Supp}(M) \cup \mathrm{Supp}(N)$),*

$$f_{\mathcal{P}} : M_{\mathcal{P}} \to N_{\mathcal{P}}$$

is injective (resp. surjective, bijective);

(iii) *for all maximal ideals $\mathcal{M} \in \mathrm{Supp}_{\mathrm{m}}(M)$ (resp. $\mathcal{M} \in \mathrm{Supp}_{\mathrm{m}}(N)$, $\mathcal{M} \in \mathrm{Supp}_{\mathrm{m}}(M) \cup \mathrm{Supp}_{\mathrm{m}}(N)$),*

$$f_{\mathcal{M}} : M_{\mathcal{M}} \to N_{\mathcal{M}}$$

is injective (resp. surjective, bijective).

Proof Let $K = \ker f$. Note that $\mathrm{Supp}(K) \subset \mathrm{Supp}(M)$. If $f_{\mathcal{M}}$ is injective for all $\mathcal{M} \in \mathrm{Supp}_{\mathrm{m}}(M)$, then $K_{\mathcal{M}} = (0)$ for all $\mathcal{M} \in \mathrm{Supp}_{\mathrm{m}}(K)$, hence $K = (0)$ and f is injective.

One studies $\mathrm{coker}\, f$ in the same way.

We have proved (iii) $\Rightarrow$ (i), the only non-trivial implication of the corollary. □

Proposition 7.29 *Let M be an A-module. Then $\mathcal{P} \in \mathrm{Supp}(M)$ implies $((0) : M) \subset \mathcal{P}$. Conversely if M is finitely generated, then $((0) : M) \subset \mathcal{P}$ implies $\mathcal{P} \in \mathrm{Supp}(M)$.*

Proof Note first that if $((0) : M) \not\subset \mathcal{P}$, there exists $s \in (A \setminus \mathcal{P}) \cap ((0) : M)$. This shows $M = \ker[M \to M_{\mathcal{P}}]$, hence $M_{\mathcal{P}} = (0)$, and $\mathcal{P} \notin \mathrm{Supp}(M)$.

Let $(x_1, ..., x_n)$ be a finite system of generators of M. We have $((0) : M) = \cap_i (0 : x_i)$. If $((0) : M) \subset \mathcal{P}$, there exists i such that $(0 : x_i) \subset \mathcal{P}$. This shows $s x_i \neq 0$ for all $s \notin \mathcal{P}$, in other words $x_i \notin \ker[M \to M_{\mathcal{P}}]$ and $x_i/1 \neq 0 \in M_{\mathcal{P}}$. □

Corollary 7.30 *If M is a finitely generated A-module, then $\mathrm{Supp}(M)$ is a closed set of $\mathrm{Spec}(A)$ for the Zariski topology.*

Proof The set defined by the ideal $((0):M)$ is closed. $\qquad\square$

Exercise 7.31 Show that the support of the $\mathbb{Z}$-module $\mathbb{Q}/\mathbb{Z}$ is not a closed set of $\mathrm{Spec}(\mathbb{Z})$ for the Zariski topology.

Definition 7.32 *A finitely generated A-module M such that $M_{\mathcal{P}}$ is a free $A_{\mathcal{P}}$-module for all $\mathcal{P} \in \mathrm{Spec}(A)$ is called locally free. If $\mathrm{rk}(M_{\mathcal{P}}) = r$ for all $\mathcal{P} \in \mathrm{Spec}(A)$, we say that M is locally free of rank r.*

Note that if M is locally free of positive rank r, then $\mathrm{Supp}(M) = \mathrm{Spec}(A)$. Note furthermore that in this case the $A_{\mathcal{P}}/\mathcal{P}A_{\mathcal{P}}$-vector space $M_{\mathcal{P}}/\mathcal{P}M_{\mathcal{P}}$ has rank r for all $\mathcal{P} \in \mathrm{Spec}(A)$.

Theorem 7.33 *Let M be a finitely generated A-module. The function*

$$\mathrm{Spec}(A) \to \mathbb{N}, \quad \mathcal{P} \to \mathrm{rk}_{(A_{\mathcal{P}}/\mathcal{P}A_{\mathcal{P}})}(M_{\mathcal{P}}/\mathcal{P}M_{\mathcal{P}}),$$

is upper semi-continuous.

Proof Let $\mathcal{P} \in \mathrm{Spec}(A)$ and $r = \mathrm{rk}_{(A_{\mathcal{P}}/\mathcal{P}A_{\mathcal{P}})}(M_{\mathcal{P}}/\mathcal{P}M_{\mathcal{P}})$. Firstly there exist $x_1, ..., x_r \in M$ such that

$$\mathrm{cl}(x_1/1), ..., \mathrm{cl}(x_r/1) \in M_{\mathcal{P}}/\mathcal{P}M_{\mathcal{P}}$$

generate $M_{\mathcal{P}}/\mathcal{P}M_{\mathcal{P}}$. By Nakayama's lemma, the elements $x_1/1, ..., x_r/1 \in M_{\mathcal{P}}$ generate this $A_{\mathcal{P}}$-module. Consider N the submodule of M generated by $x_1, ..., x_r$ and the exact sequence

$$0 \to N \to M \to M/N \to 0.$$

We assumed $(M/N)_{\mathcal{P}} = (0)$. By Proposition 7.22, there exists $s \notin \mathcal{P}$ such that $(M/N)_s = (0)$. In other words, for all prime ideals $\mathcal{Q}$ contained in the open neighbourhood $D(s) = \mathrm{Spec}(A) - V(sA)$ of $\mathcal{P}$ in $\mathrm{Spec}(A)$, we have $N_{\mathcal{Q}} = M_{\mathcal{Q}}$.

This shows that $x_1/1, ..., x_r/1 \in M_{\mathcal{Q}}$ generate this $A_{\mathcal{Q}}$-module. As an obvious consequence, we see that $\mathrm{cl}(x_1/1), ..., \mathrm{cl}(x_r/1) \in M_{\mathcal{Q}}/\mathcal{Q}M_{\mathcal{Q}}$ generate this $A_{\mathcal{Q}}/\mathcal{Q}A_{\mathcal{Q}}$-vector space, whose rank is therefore at most r, and we are done. $\qquad\square$

Example 7.34 If $M = A/\mathcal{I}$, we have

$$\mathrm{rk}_{(A_{\mathcal{P}}/\mathcal{P}A_{\mathcal{P}})}(M_{\mathcal{P}}/\mathcal{P}M_{\mathcal{P}}) = 1 \iff \mathcal{I} \subset \mathcal{P} \iff \mathcal{P} \in \mathrm{Supp}(M)$$

and

$$\mathrm{rk}_{(A_{\mathcal{P}}/\mathcal{P}A_{\mathcal{P}})}(M_{\mathcal{P}}/\mathcal{P}M_{\mathcal{P}}) = 0 \iff \mathcal{I} \not\subset \mathcal{P} \iff \mathcal{P} \notin \mathrm{Supp}(M).$$

Exercise 7.35 Let A be a domain and M a finitely generated A-module. Assume that $rk_{(A_\mathcal{P}/\mathcal{P}A_\mathcal{P})}(M_\mathcal{P}/\mathcal{P}M_\mathcal{P}) = r$ for all $\mathcal{P} \in \operatorname{Spec}(A)$. Show that M is locally free of rank r.

Theorem 7.36 *Let M be a finitely generated module on a Noetherian ring A. Then M is of finite length if and only if all prime ideals in $\operatorname{Supp}(M)$ are maximal.*

Proof If M has finite length, let $(0) = M_0 \subset M_1 \subset ... \subset M_l = M$ be a composition series of M. For all $i \geq 1$, there is a maximal ideal $\mathcal{M}_i$ such that $M_i/M_{i-1} \simeq A/\mathcal{M}_i$. We have

$$\operatorname{Supp}(M) = \bigcup_i \operatorname{Supp}(M_i/M_{i-1}) = \{\mathcal{M}_1, ..., \mathcal{M}_l\}.$$

Conversely, we know from Theorem 5.9, that there exist prime ideals $\mathcal{P}_i$ and an increasing sequence $(0) = M_0 \subset M_1 \subset ... \subset M_l = M$ of submodules of M such that $M_i/M_{i-1} \simeq A/\mathcal{P}_i$. Clearly $\mathcal{P}_i \in \operatorname{Supp}(M)$. If $\mathcal{P}_i$ is maximal for all i, it is a composition series of M, which is therefore of finite length.   □

Our next result will be particularly useful when studying quasi-coherent sheaves. We believe nevertheless that this is the right time to state it and prove it.

Theorem 7.37 *Let M be an A-module.*

(i) *The natural application*

$$f : M \to \prod_{\mathcal{M} \in \operatorname{Supp}_m(M)} M_\mathcal{M}, \quad f(x) = (x/1)_{\mathcal{M} \in \operatorname{Supp}_m(M)}$$

is injective.

(ii) *If M is finitely generated or if A is Noetherian, then $f(M)$ is the module formed by all $(x_\mathcal{M}/s_\mathcal{M})_{\mathcal{M} \in \operatorname{Supp}_m(M)}$ such that $x_\mathcal{M} s_{\mathcal{M}'} = x_{\mathcal{M}'} s_\mathcal{M}$ for all $\mathcal{M}, \mathcal{M}' \in \operatorname{Supp}_m(M)$.*

(iii) *If $\operatorname{Supp}(M)$ is finite and contains only maximal ideals, then*

$$f(M) = \prod_{\mathcal{M} \in \operatorname{Supp}(M)} M_\mathcal{M}.$$

Proof (i) Consider $x \in \ker(f)$. We have $(Ax)_{\mathcal{M}} = (0)$ for all $\mathcal{M} \in \text{Supp}_{\text{m}}(M)$. Since $\text{Supp}(Ax) \subset \text{Supp}(M)$ we have proved $(Ax)_{\mathcal{M}} = (0)$ for all $\mathcal{M} = \text{Supp}_{\text{m}}(Ax)$, hence $x = 0$ by Proposition 7.27.

(ii) We denote by M' the submodule of $\prod_{\mathcal{M} \in \text{Supp}_{\text{m}}(M)} M_{\mathcal{M}}$ formed by all $(x_{\mathcal{M}}/s_{\mathcal{M}})_{\mathcal{M} \in \text{Supp}_{\text{m}}(M)}$ such that $x_{\mathcal{M}} s_{\mathcal{M}'} = x_{\mathcal{M}'} s_{\mathcal{M}}$ for all $\mathcal{M}, \mathcal{M}' \in \text{Supp}_{\text{m}}(M)$.

It is clear that $f(M) \subset M'$. Note that $(0) : M = (0) : M'$. If M is finitely generated, this shows $\text{Supp}(M') \subset \text{Supp}(M)$ by Proposition 7.29, hence $\text{Supp}(M') = \text{Supp}(M)$.

Consider an element $x = (x_{\mathcal{M}}/s_{\mathcal{M}})_{\mathcal{M}} \in M'$. We have

$$s_{\mathcal{M}} x_{\mathcal{M}'}/s_{\mathcal{M}'} = x_{\mathcal{M}}/1 \in M_{\mathcal{M}'} \text{ for all } \mathcal{M}' \in \text{Supp}_{\text{m}}(M).$$

This shows $s_{\mathcal{M}} x \in f(M)$, hence $(f(M)_{\mathcal{M}} = M'_{\mathcal{M}}$ for all $\mathcal{M} \in \text{Supp}_{\text{m}}(M)$.

When M is finitely generated, we have $\text{Supp}(M') = \text{Supp}(M)$ and this proves $f(M) = M'$

When M is not finitely generated, we assume that A is Noetherian. Note that $f(M) \cap Ax$ is finitely generated and put $N = f^{-1}(f(M) \cap Ax$. This is a finitely generated submodule of M such that $x_{\mathcal{M}} \in N$ for all $\mathcal{M} \in \text{Supp}_{\text{m}}(M)$. This proves $x \in f(N)$ and we are done.

(iii) Assume now that $\text{Supp}(M)$ is finite and contains only maximal ideals. Let $\mathcal{N}, \mathcal{M} \in \text{Supp}(M) = \text{Supp}_{\text{m}}(M)$. We claim that $(M_{\mathcal{M}})_{\mathcal{N}} = (0)$ if $\mathcal{N} \neq \mathcal{M}$. Indeed, there exist $s \notin \mathcal{N}$ and $t \notin \mathcal{M}$ such that $st \in \mathcal{P}$ for all $\mathcal{P} \in \text{Supp}_{\text{m}}(M)$. If $x \in M$, we have $st \in \sqrt{0 : x}$. This shows $(Ax)_{st} = 0$, hence $((Ax)_{\mathcal{M}})_{\mathcal{N}} = (0)$ and $(M_{\mathcal{M}})_{\mathcal{N}} = (0)$.

Now since $\text{Supp}_{\text{m}}(M)$ is finite, we have by Proposition 7.17

$$\left(\prod_{\mathcal{M} \in \text{Supp}_{\text{m}}(M)} M_{\mathcal{M}} \right)_{\mathcal{N}} = \prod_{\mathcal{M} \in \text{Supp}_{\text{m}}(M)} (M_{\mathcal{M}})_{\mathcal{N}} = M_{\mathcal{N}}$$

and

$$f_{\mathcal{N}} : M_{\mathcal{N}} \to \left(\prod_{\mathcal{M} \in \text{Supp}(M)} M_{\mathcal{M}} \right)_{\mathcal{N}} = M_{\mathcal{N}}$$

is an isomorphism for all $\mathcal{N} \in \text{Supp}_{\text{m}}(M)$. By Corollary 7.28, f is an isomorphism. $\square$

Corollary 7.38 *Let A be a Noetherian ring with only finitely many maximal ideals. If M is a finitely generated A-module such that $M_{\mathcal{M}}$ is a free $A_{\mathcal{M}}$-module of rank n for all maximal ideals $\mathcal{M}$ of A, then M is a free A-module of rank n.*

Proof Let $\mathcal{M}_i$, $i = 1, ..., r$, be the maximal ideals of A.

Assume first that A is Artinian. Then $A \simeq \prod_i A_{\mathcal{M}_i}$ and $M \simeq \prod_i M_{\mathcal{M}_i}$. For $i = 1, ..., r$, let $e_{i,j}$, with $j = 1, ..., n$, be a basis of the free $A_{\mathcal{M}_i}$-module $M_{\mathcal{M}_i}$. Then it is easy to check that $e_j = \prod_{i=1}^{r} e_{i,j}$ is a basis of M.

If A is not Artinian, put $\mathcal{I} = \cap_i \mathcal{M}_i$. Then $B = A/\mathcal{I}$ is obviously an Artinian ring and $N = M/\mathcal{I}M$ a finitely generated B-module such that $N_\mathcal{M}$ is a free $B_\mathcal{M}$-module of rank n for all maximal ideals $\mathcal{M}$ of B. By the preceding case, N is free of rank n, hence there exists an isomorphim $nB \simeq N$.

Consider the composed surjective homomorphism $g : nA \to nB \simeq N$. We claim that it factorizes through an isomorphism $nA \simeq M$.

Let (e_i), $i = 1, ..., n$, be the canonical basis of nA. Choose elements $x_i \in M$ such that $g(e_i) = \text{cl}(x_i) \in N$ and define $f : nA \to M$ by $f(e_i) = x_i$. Note first that since $\mathcal{I}$ is the Jacobson radical of A, the elements x_i generate M by Nakayama's lemma. Hence f is surjective. As a consequence $f_{\mathcal{M}_i} : nA_{\mathcal{M}_i} \to M_{\mathcal{M}_i}$ is surjective, for all i. By Proposition 2.32, this shows that $f_{\mathfrak{M}_i}$ is an isomorphism, for all i. Finally f is an isomorphism by Corollary 7.28.

$\square$

We conclude this section with generalizations of Theorem 1.62. Note that two ideals $\mathcal{I}$ and $\mathcal{J}$ of a ring A are comaximal if and only if $\text{Supp}(A/\mathcal{I}) \cap \text{Supp}(A/\mathcal{J}) = \emptyset$.

Proposition 7.39 *Let M be an A-module. If K_i, with $i = 1, ..., n$, are submodules of M such that $\text{Supp}(M/K_i) \cap \text{Supp}(M/K_j) = \emptyset$ for $i \neq j$, the natural injective homomorphism*

$$f : M/(\bigcap_i K_i) \to \bigoplus_i (M/K_i)$$

is an isomorphism.

Proof Since f is injective, we have $\text{Supp}(M/(\cap_i K_i)) \subset \bigcup_i \text{Supp}(M/K_i)$. But $\text{Supp}(M/K_i) \subset \text{Supp}(M/(\cap_i K_i))$ is obvious and we find $\text{Supp}(M/(\cap_i K_i)) = \cup_i \text{Supp}(M/K_i)$. Consider $\mathcal{P} \in \text{Supp}(M/(\cap_i K_i))$. There is a unique integer j such that $\mathcal{P} \in \text{Supp}(M/K_j)$. As a consequence, the homomorphism

$$f_\mathcal{P} : (M/(\bigcap_i K_i))_\mathcal{P} \to (\bigoplus_i (M/K_i))_\mathcal{P} = (M/K_j)_\mathcal{P}$$

is surjective, hence f is surjective by Corollary 7.28. $\square$

Exercise 7.40 Assume $\text{Supp}(M) = \{\mathcal{M}_1, ..., \mathcal{M}_n\}$, where $\mathcal{M}_i$ is maximal for all i. If, for all i, we put $K_i = \ker(M \to M_{\mathcal{M}_i})$, show that

$$M_{\mathcal{M}_i} = M/K_i.$$

When the ring is Noetherian, Proposition 7.39 can be improved in the following way:

Proposition 7.41 *Let A be a Noetherian ring and M be an A-module. Assume there exists a finite number of closed sets $F_i \subset \mathrm{Supp}(M) \subset \mathrm{Spec}(A)$ such that*

$$F_i \cap F_j = \emptyset \quad \text{for } i \neq j \quad \text{and} \quad \mathrm{Supp}(M) = \bigcup_{i=1}^{n} F_i.$$

If $T_i(M) = \{x \in M, \ \mathrm{Supp}(Ax) \subset F_i\}$, then $M = \oplus_{i=1}^{n} T_i(M)$.

Proof To begin with, note that $\mathrm{Supp}(\sum_{j \neq i} T_j(M)) \subset \cup_{j \neq i} F_j$. This implies

$$\mathrm{Supp}(T_i(M)) \cap \mathrm{Supp}(\sum_{j \neq i} T_j(M)) = \emptyset.$$

As a consequence, we get

$$T_i(M) \cap (\sum_{j \neq i} T_j(M)) = (0) \quad \text{and} \quad \sum_{i=1}^{n} T_i(M) \simeq \bigoplus_{i=1}^{n} T_i(M).$$

To show $M = \sum_{i=1}^{n} T_i(M)$, consider $x \in M$ and $\mathcal{I} = ((0) : x)$. We recall that $Ax \simeq A/\mathcal{I}$. Next, consider a minimal primary decomposition $\mathcal{I} = \cap_{i=1}^{r} \mathcal{Q}_i$. If $\mathcal{P}_s$ is the radical of $\mathcal{Q}_s$, there exists $j \in [1, n]$ such that $\mathcal{P}_s \in F_j$, hence $V(\mathcal{Q}_s) = V(\mathcal{P}_s) \subset F_j$. Let $E_i \subset [1, n]$ be the set of integers $s \in [1, n]$ such that $V(\mathcal{Q}_s) \subset F_i$ and put $\mathcal{I}_i = \cap_{i \in E_i} \mathcal{Q}_s$. It is clear that $V(\mathcal{I}_i) \subset F_i$, hence that the ideals $\mathcal{I}_i$ are pairwise comaximal, and that $\mathcal{I} = \cap_{i=1}^{n} \mathcal{I}_i$. By Theorem 1.62, there is a natural isomorphism $A/\mathcal{I} \simeq \oplus_{i=1}^{n} A/\mathcal{I}_i$. This isomorphism induces obviously an isomorphism $T_i(Ax) \simeq T_i(A/\mathcal{I}) \simeq A/\mathcal{I}_i$ and we are done.   □

7.4   Localization of ideals

If $\mathcal{I}$ is an ideal of A, we have seen that $S^{-1}\mathcal{I}$ is a submodule of $S^{-1}A$, hence an ideal of $S^{-1}A$. We often denote this ideal by $\mathcal{I}S^{-1}A$. Clearly, we have

$$\mathcal{I}S^{-1}A = \{a/s\} \quad \text{with} \quad a \in \mathcal{I} \quad \text{and} \quad s \in S,$$

$$\mathcal{I} \cap S \neq \emptyset \iff \mathcal{I}S^{-1}A = S^{-1}A,$$

$$(\mathcal{I} \cap \mathcal{I}')S^{-1}A = \mathcal{I}S^{-1}A \cap \mathcal{I}'S^{-1}A \quad \text{and} \quad (\mathcal{I} + \mathcal{I}')S^{-1}A = \mathcal{I}S^{-1}A + \mathcal{I}'S^{-1}A.$$

Let $\mathcal{J}$ be an ideal of $S^{-1}A$. We denote, as in Proposition 7.17, by $\mathcal{J} \cap A$ the inverse image $l_A^{-1}(\mathcal{J})$ of $\mathcal{J}$ by the localization homomorphism $l_A : A \to S^{-1}A$. Clearly, $\mathcal{J} \cap A$ is an ideal of A.

We have already seen, in the more general case of modules (Proposition 7.17), the following result.

Proposition 7.42 *For all ideals $\mathcal{J}$ of $S^{-1}A$, we have*

$$(\mathcal{J} \cap A)S^{-1}A = \mathcal{J}.$$

As an immediate consequence of this last result, we get

Corollary 7.43 *If A is a principal ideal ring and S a multiplicatively closed part of A, then $S^{-1}A$ is a principal ideal ring.*

Proposition 7.44 *If $\mathcal{I}$ is an ideal of A, then*

$$\mathcal{I}S^{-1}A \cap A = \{b \in A, \quad \text{there exists } s \in S \text{ such that } sb \in \mathcal{I}\}.$$

Proof We know that $b \in \mathcal{I}S^{-1}A \cap A$ if and only if there exist $a \in \mathcal{I}$ and $t \in S$ such that $b/1 = a/t \in S^{-1}A$. But this is equivalent to the existence of $s \in S$ such that $sb \in \mathcal{I}$ and we are done. $\qquad\square$

Corollary 7.45 *Let S be a multiplicatively closed part and $\mathcal{I}$ an ideal of the ring A. Then we have $\mathcal{I}S^{-1}A \cap A = \mathcal{I}$ if and only if $\mathrm{cl}(s) \in A/\mathcal{I}$ is not a zero divisor for all $s \in S$.*

Proof Since $b \in \mathcal{I}S^{-1}A \cap A$ if and only if there exists $s \in S$ such that $\mathrm{cl}(s)\mathrm{cl}(b) = 0 \in A/\mathcal{I}$, this is clear. $\qquad\square$

As an obvious consequence we get

Corollary 7.46 *Let $\mathcal{P}$ be a prime ideal of A. If $\mathcal{P} \cap S = \emptyset$, then $\mathcal{P}S^{-1}A$ is a prime ideal of $S^{-1}A$ and $\mathcal{P}S^{-1}A \cap A = \mathcal{P}$.*

Using Proposition 7.42 and Corollary 7.46, we immediately get

Theorem 7.47 *The maps*

$$\mathcal{P} \to \mathcal{P}S^{-1}A \text{ and } \mathcal{P}' \to \mathcal{P}' \cap A$$

define a bijective correspondence between the set of prime ideals of A disjoint from S and the set of prime ideals of $S^{-1}A$.

As special cases of this theorem, we find the two following corollaries.

Corollary 7.48 *If $\mathcal{P}$ is a prime ideal of A, then $\mathcal{P}A_{\mathcal{P}}$ is the unique maximal ideal of the ring $A_{\mathcal{P}}$, in other words $A_{\mathcal{P}}$ is a local ring with maximal ideal $\mathcal{P}A_{\mathcal{P}}$.*

Corollary 7.49 *If $s \in A$, the map $\mathcal{P}' \to \mathcal{P}' \cap A$ is a homeomorphism from* $\operatorname{Spec}(A_s)$ *to the open set* $D(s) = \operatorname{Spec}(A) \setminus V(sA)$ *of* $\operatorname{Spec}(A)$.

Exercise 7.50 Let A be a ring and $\mathcal{P}_i$, with $i = 1, \ldots, n$, be prime ideals of A such that $\mathcal{P}_i \not\subset \mathcal{P}_j$ for $i \neq j$. If $S = \bigcap_{i=1}^{n}(A \setminus \mathcal{P}_i)$, show that the maximal ideals of $S^{-1}A$ are $\mathcal{P}_i S^{-1}A$, with $i = 1, \ldots, n$.

We end this section with a useful technical lemma.

Lemma 7.51 *Let A be a subring of B. Let S be a multiplicatively closed part of A and $\mathcal{P}$ a prime ideal of A disjoint from S. If $\mathcal{N}$ is a prime ideal of B disjoint from S,*

$$\mathcal{N} \cap A = \mathcal{P} \iff \mathcal{N}S^{-1}B \cap S^{-1}A = \mathcal{P}S^{-1}A.$$

Proof Assume $\mathcal{N} \cap A = \mathcal{P}$. Then $(\mathcal{N}S^{-1}B \cap B) \cap A = \mathcal{P}$, hence $(\mathcal{N}S^{-1}B \cap S^{-1}A) \cap A = \mathcal{P}$. This implies $\mathcal{N}S^{-1}B \cap S^{-1}A = \mathcal{P}S^{-1}A$ by Theorem 7.47. Conversely, if $\mathcal{N}S^{-1}B \cap S^{-1}A = \mathcal{P}S^{-1}A$, we have $\mathcal{N}S^{-1}B \cap S^{-1}A \cap A = \mathcal{P}$. This proves $\mathcal{N}S^{-1}B \cap B \cap A = \mathcal{P}$, that is $\mathcal{N} \cap A = \mathcal{P}$ by twice using Theorem 7.47. $\square$

7.5   Localization and UFDs

The following lemma is a convenient characterization of UFDs. Its proof is straightforward and left to the reader.

Lemma 7.52 *Let A be a domain and $a_i \in A$, with $i \in E$, be elements satisfying the following conditions.*

(i) *The ideal $a_i A$ is a prime ideal for all $i \in E$.*

(ii) *Every non-zero element A is a product of some of the a_i and a unit of A.*

Then A is a UFD and for each irreducible element $b \in A$, there exists $i \in E$ such that $bA = a_i A$.

Theorem 7.53 *If A is a UFD and S a multiplicatively closed part of A, then $S^{-1}A$ is a UFD.*

Proof Let $a \in A$ be irreducible. If $aA \cap S = \emptyset$, then $aS^{-1}A$ is a prime ideal of $S^{-1}A$, hence $a/1 \in S^{-1}A$ is irreducible. We claim that every non-zero element of $S^{-1}A$ is a product of irreducible elements of this form and a unit, hence that $S^{-1}A$ is a UFD by Lemma 7.52.

Note that if $aA \cap S \neq \emptyset$, then $a/1 \in S^{-1}A$ is obviously a unit.

Consider $b/s \in S^{-1}A$. If $b = a_1...a_n$ is a decomposition as a product of irreducible elements, then $b/1 = (a_1/1)...(a_n/1)$ is a decomposition as a product of irreducible elements and units. Then $b/s = (a_1/1)...(a_n/1)(1/s)$ and we are done, since $1/s$ is also a unit. $\square$

Theorem 7.54 *Let A be a domain and s_i, with $i \in E$, non-zero elements of A satisfying the following conditions:*

(i) *the ideals $s_i A$ are prime for all $i \in E$;*

(ii) *if S is the multiplicatively closed part of A formed by finite products of s_i, then for every non-zero element $b \in A$ there exist $s \in S$ and $b' \in A$ such that $b = sb'$ and $b' \notin s_i A$ for all $i \in E$.*

If $S^{-1}A$ is a UFD, so is A. Furthermore, $a \in A$ is irreducible if and only if one of the following conditions is satisfied:

(1) there exists $i \in E$ such that $aA = s_i A$;
(2) the element $a/1 \in S^{-1}A$ is irreducible and $a \notin s_i A$ for all $i \in E$.

Our proof depends on the following little lemma.

Lemma 7.55 *If $a \notin s_i A$ for all $i \in E$ then $aA \cap sA = asA$ for all $s \in S$.*

Proof Assume $ab = sc$. Put $s = s_1^{n_1}...s_r^{n_r}$. Since $s_1 A$ is a prime ideal and $a \notin s_1 A$, there exists $b' \in A$ such that $b = s_1 b'$. This shows $ab' = s_1^{n_1-1}...s_r^{n_r}c$. We are done by an obvious induction on $\sum n_i$. $\square$

Proof of 7.54

Consider an irreducible element $b/t \in S^{-1}A$. The ideal $(b/t)S^{-1}A = bS^{-1}A$ is prime. If $b = sb'$ where $s \in S$ and $b' \notin s_i A$ for all $i \in E$, we claim that $b'A = bS^{-1}A \cap A$. Indeed, if $z \in bS^{-1}A \cap A = b'S^{-1}A \cap A$, there exists $s' \in S$ such that $s'z \in b'A$. By our little lemma, this shows $s'z \in b's'A$, hence $z \in b'A$. Note that this implies also $b'S^{-1}A = (b/t)S^{-1}A$.

If $x \in A$ is non-zero, we have proved the existence, in the fraction ring $S^{-1}A$, of a decomposition $x/1 = (a/t)(b_1/1)...(b_n/1)$ such that

(i) the element a/t is a unit in $S^{-1}A$,

(ii) for $j = 1, ..., n$ the ideal $b_j A$ is prime and $b_j \notin s_i A$ for all $i \in E$.

Note that $b_1...b_n \notin s_i A$ for all $i \in E$. By using once more our favourite little lemma, we find $tx = ab_1...b_n \in tA \cap b_1...b_n A = tb_1...b_n A$, hence $x = cb_1...b_n$. This shows $c/1 = a/t \in S^{-1}A$. Consequently $c/1 \in S^{-1}A$ is a unit. We recall that there exists $s \in S$ and $c' \in A$ such that $c = sc'$ and $c' \notin s_i A$ for all

$i \in E$. Now we claim that c' is a unit of A. Indeed, if $y/s' \in S^{-1}A$ is such that $(c'/1)(y/s') = 1$, then $c'y = s' \in c'A \cap s'A = c's'A$ (what a wonderful little lemma!), hence $y \in s'A$ and c' is a unit.

We have therefore proved $x = c'sb_1...b_n$, where c' is a unit, $s \in S$ and b_jA is a prime ideal for all j. Since s is a product of elements s_i, $i \in E$, we have proved Theorem 7.54, using Lemma 7.52.   □

Remark 7.56 *If the ring A is Noetherian and s_i, with $i \in E$, is a set of elements of A such that s_iA is a prime ideal for all $i \in E$, then condition (ii) in our theorem is automatically satisfied.*

If not, consider the set of principal ideals bA such that $b \in A$ does not satisfy condition (ii). Let cA be maximal in this set. There must be an $i \in E$ such that $c \in s_iA$. If $c = s_id$, then cA is strictly contained in dA, hence d satisfies condition (ii). There is a decomposition $d = sd'$, with $s \in S$ and $d' \notin s_iA$ for all $i \in E$. But the decomposition $c = s_isd'$ shows that c satisfies condition (ii). This is a contradiction.

As an immediate consequence of Theorem 7.54, we get Theorem 1.72, whose proof was postponed.

Corollary 7.57 *If R is a UFD, then so is the polynomial ring $R[X]$. The irreducible elements of $R[X]$ are*

(i) *the irreducible elements of R,*

(ii) *the polynomials $P \in R[X]$ whose coefficients have gcd 1 and such that P is irreducible in $K(R)[X]$, where $K(R)$ is the fraction field of R.*

Proof In the ring $A = R[X]$ consider the irreducible elements s_i of R. The multiplicatively closed part generated by these elements is $S = R - \{0\}$. If $P \in A = R[X]$, let s be the gcd of its coefficients and put $P' = P/s$. It is clear that $P' \notin s_iR[X]$ for all i. Since $S^{-1}A = K(R)[X]$ is a UFD, we can apply Theorem 7.54.   □

7.6   Localization and primary decomposition

Proper ideals in Noetherian rings have a primary decomposition. Fraction rings of Noetherian rings are Noetherian. Let $\mathcal{I}$ be an ideal of a Noetherian ring A such that $\mathcal{I}S^{-1}A \neq S^{-1}A$. In this section we explain how we can relate the primary decompositions of $\mathcal{I}$ (in A), of $\mathcal{I}S^{-1}A$ (in $S^{-1}A$) and of $\mathcal{I}S^{-1}A \cap A$ (in A).

Theorem 7.58 *Let A be a Noetherian ring and S a multiplicatively closed part of A.*

(i) *If $\mathcal{Q}$ is a $\mathcal{P}$-primary ideal of A such that $\mathcal{Q} \cap S = \emptyset$, then $\mathcal{P} \cap S = \emptyset$ and $\mathcal{Q}S^{-1}A$ is a $\mathcal{P}S^{-1}A$-primary ideal of $S^{-1}A$ such that $\mathcal{Q}S^{-1}A \cap A = \mathcal{Q}$.*

(ii) *If $\mathcal{Q}'$ is a $\mathcal{P}'$-primary ideal of $S^{-1}A$, then $(\mathcal{Q}' \cap A)$ is a $(\mathcal{P}' \cap A)$-primary ideal of A disjoint from S and $(\mathcal{Q}' \cap A)S^{-1}A = \mathcal{Q}'$.*

We have thus defined a bijective correspondence between the set of primary ideals of A disjoint from S and the set of primary ideals of $S^{-1}A$.

Proof As for the case of prime ideals, this result is a straightforward consequence of Proposition 7.42 and Corollary 7.46, with the help of the following easy equivalences.

Let $\mathcal{Q}$ be a primary ideal of A and $\mathcal{P} = \sqrt{\mathcal{Q}}$, then

$$\mathcal{Q} \cap S = \emptyset \Leftrightarrow \mathcal{P} \cap S = \emptyset \Leftrightarrow s \in S \text{ and } sx \in \mathcal{Q} \Rightarrow x \in \mathcal{Q}.$$

$\square$

Corollary 7.59 *Let A be a Noetherian ring and S a multiplicatively closed part of A. If $\mathcal{I}$ is an ideal of A and $\mathcal{I} = \cap_1^n \mathcal{Q}_i$ a minimal primary decomposition of $\mathcal{I}$, then*

(i) *$\mathcal{I}S^{-1}A = \cap_{(\mathcal{Q}_i \cap S = \emptyset)} \mathcal{Q}_i S^{-1}A$. If $\mathcal{I}S^{-1}A \neq S^{-1}A$, this is a minimal primary decomposition of $\mathcal{I}S^{-1}A$ in $S^{-1}A$;*

(ii) *$\mathcal{I}S^{-1}A \cap A = \cap_{(\mathcal{Q}_i \cap S = \emptyset)} \mathcal{Q}_i$. If $\mathcal{I}S^{-1}A \cap A \neq A$, this is a minimal primary decomposition of $\mathcal{I}S^{-1}A \cap A$ in A.*

Proof If $\mathcal{Q}_i \cap S \neq \emptyset$, then $\mathcal{Q}_i S^{-1}A = S^{-1}A$. If not, $\mathcal{Q}_i S^{-1}A$ is a primary ideal and $\mathcal{Q}_i S^{-1}A \cap A = \mathcal{Q}_i$. This proves (i). For (ii) we need also the following general lemma.

Lemma 7.60 *If $\mathcal{I}$ and $\mathcal{J}$ are ideals of A, then*

$$(\mathcal{I} \cap \mathcal{J})S^{-1}A \cap A = (\mathcal{I}S^{-1}A \cap A) \cap (\mathcal{J}S^{-1}A \cap A).$$

Proof One inclusion is obvious. For the other, consider $a \in \mathcal{I}S^{-1}A \cap A \cap \mathcal{J}S^{-1}A \cap A$. There exist $s \in S$ such that $sa \in \mathcal{I}$ and $t \in S$ such that $ta \in \mathcal{J}$. This shows $sta \in \mathcal{I} \cap \mathcal{J}$, hence $a \in (\mathcal{I} \cap \mathcal{J})S^{-1}A \cap A$. $\square$

Corollary 7.61 *Let A be a Noetherian ring, $\mathcal{I}$ a proper ideal of A and $\mathcal{I} = \cap_1^n \mathcal{Q}_i$ a minimal primary decomposition of $\mathcal{I}$. If $\mathcal{P}_i = \sqrt{\mathcal{Q}_i}$ is a minimal prime of $\mathcal{I}$, then*

$$\mathcal{Q}_i = \mathcal{I} A_{\mathcal{P}_i} \cap A.$$

Proof This is an immediate consequence of Corollary 7.59 (ii). Since $\mathcal{P}_i$ is a minimal prime of $\mathcal{I}$, we have $\mathcal{Q}_j \not\subset \mathcal{P}_i$ for $j \neq i$. □

We note that this proves yet again our Corollary 3.33, namely that the primary component of an ideal $\mathcal{I}$ for a minimal prime ideal of $\mathcal{I}$ is uniquely defined.

7.7   Back to minimal prime ideals

Equipped now with the notion of support, we return to minimal prime ideals. First we note that if $\mathcal{I}$ is an ideal of a ring A, then a prime ideal $\mathcal{P}$ is a minimal prime ideal of $\mathcal{I}$ if and only if $\operatorname{Supp}((A/\mathcal{I})_\mathcal{P}) = \{\mathcal{P}A_\mathcal{P}\}$. This allows us to give the following definition.

Definition 7.62 *Let M be an A-module and $\mathcal{P} \in \operatorname{Supp}(M)$. We say that $\mathcal{P}$ is a minimal prime ideal of M if $\operatorname{Supp}(M_\mathcal{P}) = \{\mathcal{P}A_\mathcal{P}\}$.*

When the ring is Noetherian, the next statement can be proved by using primary decomposition. Using fractions we find a shorter proof in the general case.

Proposition 7.63 *If $\mathcal{P}$ is a minimal prime ideal of A and $a \in \mathcal{P}$, there exist an integer r and an element $s \notin \mathcal{P}$ such that $a^r s = 0$.*

Proof Consider $a/1 \in A_\mathcal{P}$. Since $\mathcal{P}A_\mathcal{P}$ is the nilradical of $A_\mathcal{P}$, this element is nilpotent. But if $(a/1)^r = 0$, there exists $s \notin \mathcal{P}$ such that $a^r s = 0$. □

Definition 7.64 *(Symbolic powers of a prime ideal)*
    Let A be a *(not necessarily Noetherian)* ring and $\mathcal{P}$ a prime ideal of A. The order n symbolic power of $\mathcal{P}$ is

$$\mathcal{P}^{(n)} = \mathcal{P}^n A_\mathcal{P} \cap A.$$

Exercises 7.65

1. If $\mathcal{M}$ is a maximal ideal, show that $\mathcal{M}^{(n)} = \mathcal{M}^n$.
2. If $\mathcal{P} = aA$ is a principal prime ideal, show that $\mathcal{P}^{(n)} = a^n A = \mathcal{P}^n$.

Since $\mathcal{P}$ is the only minimal prime ideal of $\mathcal{P}^n$, the next result is a consequence of Corollary 7.61.

Proposition 7.66 *If A is Noetherian, $\mathcal{P}^{(n)}$ is the $\mathcal{P}$-primary component of $\mathcal{P}^n$.*

7.8 Localization and associated prime ideals

We conclude this chapter with a short study of the prime ideals associated to a finitely generated fraction module, when the ring is Noetherian.

Theorem 7.67 *Let A be a Noetherian ring, S a multiplicatively closed part of A and $\mathcal{P}$ a prime ideal of A such that $\mathcal{P} \cap S = \emptyset$. If M is a finitely generated A-module, then*

$$\mathcal{P}S^{-1}A \in \mathrm{Ass}(S^{-1}M) \Longleftrightarrow \mathcal{P} \in \mathrm{Ass}(M).$$

Proof Let $\mathcal{P}$ be a prime ideal of A such that $\mathcal{P} \cap S = \emptyset$. Let $\mathcal{P} \in \mathrm{Ass}(M)$ and $x \in M$ such that $\mathcal{P} = 0 : x$. The isomorphism $A/\mathcal{P} \simeq Ax$ induces an isomorphism

$$S^{-1}A/\mathcal{P}S^{-1}A \simeq S^{-1}Ax \subset S^{-1}M,$$

hence $\mathcal{P}S^{-1}A = 0 : x$. This proves

$$\mathcal{P}S^{-1}A \in \mathrm{Ass}(S^{-1}M).$$

Conversely, let $\mathcal{P}S^{-1}A \in \mathrm{Ass}(S^{-1}M)$ and $x/s \in S^{-1}M$ such that

$$\mathcal{P}S^{-1}A = (0 : (x/s)).$$

This shows first that $\mathcal{P}S^{-1}A = 0 : (x/1)$ and then that there exists $t \in S$ such that $\mathcal{P} = (0 : tx)$. □

Corollary 7.68 *Let A be a Noetherian ring and M a finitely generated A-module.*

(i) *The minimal prime ideals of M are all associated to M.*

(ii) *M has only a finite number of minimal prime ideals.*

Proof If $\mathrm{Supp}(M_{\mathcal{P}}) = \{\mathcal{P}A_{\mathcal{P}}\}$, we have $\mathrm{Ass}(M_{\mathcal{P}}) = \{\mathcal{P}A_{\mathcal{P}}\}$, hence $\mathcal{P} \in \mathrm{Ass}(M)$. Since $\mathrm{Ass}(M)$ is finite, we are done. □

Theorem 7.69 *Let A be a Noetherian ring and M a finitely generated A-module. Assume*

$$\text{rk}_{A_{\mathcal{P}}/\mathcal{P}A_{\mathcal{P}}}(M_{\mathcal{P}}/\mathcal{P}M_{\mathcal{P}}) = n \quad \text{for all } \mathcal{P} \in \text{Spec}(A).$$

If $M_{\mathcal{P}}$ is a free $A_{\mathcal{P}}$-module for all $\mathcal{P} \in \text{Ass}(A)$, then M is locally free of rank n.

Proof From the definition of locally free modules, we can assume that A is local. Let $\mathcal{M}$ be its maximal ideal. Since $\text{rk}_{A/\mathcal{M}}(M/\mathcal{M}M) = n$, there exist $x_1, ..., x_n \in M$ generating M, by Nakayama's lemma. Let K be the kernel of the surjective homomorphism

$$f : nA \to M, \quad (a_1, ..., a_n) \to \sum a_i x_i.$$

Consider $\mathcal{P} \in \text{Ass}(A)$. Since $\text{rk}_{A_{\mathcal{P}}/\mathcal{P}A_{\mathcal{P}}}(M_{\mathcal{P}}/\mathcal{P}M_{\mathcal{P}}) = n$, the free $A_{\mathcal{P}}$-module $M_{\mathcal{P}}$ has rank n, by Nakayama's lemma. Hence the system of generators $(x_1/1, ..., x_n/1)$ of $M_{\mathcal{P}}$ is a basis by Proposition 2.32. This shows $K_{\mathcal{P}} = (0)$. Since $\text{Ass}(K) \subset \text{Ass}(nA) = \text{Ass}(A)$, we have proved $\text{Ass}(K) = \emptyset$. Hence $K = (0)$ and f is an isomorphism. □

Proposition 7.70 *Let A be a Noetherian ring and M a finitely generated, rank n, locally free A-module. There exists an injective homomorphism $f : nA \to M$ such that $f_{\mathcal{P}} : nA_{\mathcal{P}} \to M_{\mathcal{P}}$ is an isomorphism for all $\mathcal{P} \in \text{Ass}(A)$.*

Proof Let $\mathcal{P}_i$, $i = 1, ..., n$, be the prime ideals associated to A and $S = \cap_i(A \setminus \mathcal{P}_i)$. If $\mathcal{P}$ is a prime ideal such that $S \cap \mathcal{P} = \emptyset$, then $\mathcal{P} \subset \cup_i \mathcal{P}_i$. Hence there exists i such that $\mathcal{P} \subset \mathcal{P}_i$. As a consequence, $S^{-1}A$ has only finitely many maximal ideals. Since $S^{-1}M$ is a finitely generated, rank n, locally free $S^{-1}A$-module, it is free of rank n, by Corollary 7.38. Let (x_i/s_i), $i = 1, ..., n$, be a basis. Obviously, $(x_i/1)$ is also a basis. Let (e_i), $i = 1, ..., n$, be the canonical basis of nA. Consider then the homomorphism

$$f : nA \to M, \quad f(e_i) = x_i.$$

Since $S^{-1}f : nS^{-1}A \to S^{-1}M$ is an isomorphism, $f_{\mathcal{P}_i} : nA_{\mathcal{P}_i} \to M_{\mathcal{P}_i}$ is an isomorphism for all i. In particular, if $K = \ker(f)$, we have $K_{\mathcal{P}_i} = (0)$ for all i. Since $\text{Ass}(K) \subset \text{Ass}(nA) = \text{Ass}(A)$, this implies $K_{\mathcal{P}} = (0)$ for all $\mathcal{P} \in \text{Ass}(K)$, hence $K = (0)$. □

7.9 Exercises

1. Let A be a ring and $\mathcal{I}$ an ideal of A. Show that the elements which are not zero divisors modulo $\mathcal{I}$ form a multiplicatively closed part S of A. Assume furthermore that A is Noetherian and show that the ring $S^{-1}A$ has only finitely many maximal ideals.

2. Let $\mathcal{I}$ and $\mathcal{J}$ be ideals of a Noetherian ring A. Show that if $\mathcal{I}_{\mathcal{P}} \cap \mathcal{J}_{\mathcal{P}} = (0)$ for all $\mathcal{P} \in \mathrm{Ass}(A)$, then $\mathcal{I} \cap \mathcal{J} = (0)$.

3. Let A be a ring such that its Jacobson radical is (0). Let M be a finitely generated A-module and $x_1, \ldots, x_n \in M$ be elements such that for each maximal ideal $\mathcal{M}$ of A the classes $\mathrm{cl}(x_1), \ldots, \mathrm{cl}(x_n) \in M/\mathcal{M}M$ form a basis of the $A/\mathcal{M}$-vector space $M/\mathcal{M}M$. Show that M is free and that $(x_1, \ldots, x_n)$ is a basis of M.

4. Let $\mathcal{P}$ be a prime ideal in a Noetherian domain A. Show $\bigcap_{n\geq 0} \mathcal{P}^{(n)} = (0)$.

5. Let A be a Noetherian ring and $\mathcal{I}$ an ideal of A. Assume that for each $\mathcal{P} \in \mathrm{Ass}(A)$ the ideals $\mathcal{I}$ and $\mathcal{P}$ are not comaximal. Show $\bigcap_{n\geq 0} \mathcal{I}^n = (0)$. Hint: consider for each $\mathcal{P}_i \in \mathrm{Ass}(A)$ a maximal ideal $\mathcal{M}_i$ such that $\mathcal{P}_i + \mathcal{I} \subset \mathcal{M}_i$ and put $S = A \setminus (\bigcap_i \mathcal{M}_i)$. Show that S does not contain any zero divisor, that $\mathcal{I}S^{-1}A \subset \mathrm{JR}(S^{-1}A)$ and conclude.

6. Let A be a Noetherian ring and M a finitely generated A-module. Consider the Fitting ideals $F_r(M)$ as in section 5.5, exercise 3. If $\mathcal{P}$ is a prime ideal of A, show $\mathrm{rk}_{A_{\mathcal{P}}/\mathcal{P}A_{\mathcal{P}}}(M_{\mathcal{P}}/\mathcal{P}M_{\mathcal{P}}) > r$ if and only if $F_r(M) \subset \mathcal{P}$. Why is this a new proof of Theorem 7.33?

7. Let A be a Noetherian ring and $\mathcal{P}_1, \ldots, \mathcal{P}_r$ its minimal prime ideals. If M is a finitely generated A-module, put $m(M) = \sum_{i=1}^r l_{A_{\mathcal{P}_i}}(M_{\mathcal{P}_i})$. Show that m is an additive function, defined in the category of finitely generated A-modules.

8. Let A be a Noetherian ring. Consider an exact sequence of finitely generated A-modules:

$$0 \to L_r \to \ldots \to L_0 \to M \to 0.$$

Assume that L_i is free for all i. Use exercise 8 in section 5.5 to show that if $\mathcal{P} \in \mathrm{Ass}(A)$, then $M_{\mathcal{P}}$ is a free $A_{\mathcal{P}}$-module. Show furthermore that the rank of $M_{\mathcal{P}}$ does not depend on $\mathcal{P} \in \mathrm{Ass}(A)$.

8

Integral extensions of rings

If you like algebra, you will like this chapter! We have chosen to first study integral and finite extensions of rings; the special case of fields is developed later in chapter 9. In the third section the going-up theorem is easy, but the going-down is a deep and difficult result.

If A is a subring of B, we will often say that $A \subset B$ is a ring extension (or an extension when it is obvious that we are working with rings).

8.1 Algebraic elements, integral elements

Definition 8.1 *Let $A \subset B$ be a ring extension and $x \in B$.*

(i) *We say that x is algebraic over A if there exists a non-zero polynomial $P \in A[X]$ such that $P(x) = 0$.*

(ii) *If x is not algebraic over A, we say that x is transcendental over A.*

(iii) *We say that x is integral over A if there is a monic polynomial $P \in A[X]$ such that $P(x) = 0$.*

Note that an element integral over A is obviously algebraic over A. If A is a field an element algebraic over A is integral over A.

Definition 8.2 *A non-trivial relation $a_0 x^n + a_1 x^{n-1} + \ldots + a_n = 0$ with coefficients $a_i \in A$, is a relation of algebraic dependence for x over A. If $a_0 = 1$, it is a relation of integral dependence for x over A.*

Definition 8.3 *Let K be a field and x an element (in some extension of K) algebraic over K. The minimal polynomial of x over K is the unique monic polynomial of $K[X]$ which generates the kernel of the ring homomorphism $e : K[X] \to K[x]$, $e(Q(X)) = Q(x)$.*

103

We recall that $K[X]$ is a principal ideal ring, hence this definition makes sense. On the other hand, if A is not a field, $A[X]$ is not a principal ideal ring. In this case an element algebraic (or integral) over A has nothing comparable to a minimal polynomial.

Proposition 8.4 *If P is the minimal polynomial of x over K, then $\deg(P) = \mathrm{rk}_K(K[x])$.*

Proof Let $P = X^n + a_1 X^{n-1} + ... + a_n$. The isomorphism $K[X]/(P) \simeq K[x]$ shows that $(1, x, ..., x^{n-1})$ is a basis of the K-vector space $K[x]$.   □

Theorem 8.5 *Let A be a subring of B and $x \in B$. The following conditions are equivalent:*

(i) *the element x is integral over A;*

(ii) *the A-module $A[x]$ is finitely generated;*

(iii) *there exists a faithful $A[x]$-module M which is a finitely generated A-module.*

Proof (i) $\Rightarrow$ (ii). Assume x is integral over A. There is a relation $x^n + a_1 x^{n-1} + ... + a_{n-1} x + a_n = 0$, with coefficients $a_i \in A$. This relation proves that x^n is contained in the sub-A-module of B generated by $1, x, ..., x^{n-1}$. The relations
$$x^{n+m} + a_1 x^{n+m-1} + ... + a_{n-1} x^{m+1} + a_n x^m = 0,$$
prove that x^{n+m} is contained in the sub-A-module of B generated by
$$1, x, ..., x^{n+m-1}$$
and a straightforward induction on m shows that x^m is contained in the sub-A-module of B generated by $1, x, ..., x^{n-1}$, for all $m \geq 0$.

(ii) $\Rightarrow$ (iii) is obvious with $M = A[x]$.

To show (iii) $\Rightarrow$ (i), we use Cayley–Hamilton revisited (Theorem 2.44).

Let $(z_1, ..., z_n)$ be a system of generators of M as an A-module. Consider $a_{ij} \in A$ such that $x z_i = \sum_1^n a_{ij} x_j$.

Consider the monic polynomial $P(X) = \det(X I_{n \times n} - (a_{ij})) \in A[X]$. Multiplying by x in M is an endomorphism of the A-module. We know by the Cayley–Hamilton theorem that the endomorphism $P(x)$ of M is zero. But this endomorphism is the multiplication by $P(x)$ in M. Since M is a faithful $A[x]$-module, this shows $P(x) = 0$, hence x is integral over A.   □

Corollary 8.6 *If x is integral over A and $y \in A[x]$, then y is integral over A.*

Proof Obviously $A[x]$ is a faithful $A[y]$-module (use for example $1 \in A[x]$). Since it is a finitely generated A-module, we are done.   □

While proving Theorem 8.5, we have almost proved the following useful and more "precise" result.

Proposition 8.7 *Let $\mathcal{I} \subset A$ be an ideal. The following conditions are equivalent:*

(i) *there exists a relation $x^n + b_1 x^{n-1} + ... + b_n = 0$, with $b_i \in \mathcal{I}$ for all i;*

(ii) *the A-module $A[x]$ is finitely generated and there exists an integer m such that $x^m A[x] \subset \mathcal{I}A[x]$;*

(iii) *there exist a faithful $A[x]$-module M which is a finitely generated A-module and an integer m such that $x^m M \subset IM$.*

Proof (i) $\Rightarrow$ (ii) $\Rightarrow$ (iii) are proved as in Theorem 8.5. For (iii) $\Rightarrow$ (i), it is enough to find a relation $x^{mn} + b_1 x^{m(n-1)} + ... + b_n = 0$, with $b_i \in \mathcal{I}$ for all i. Hence we can assume $xM \subset \mathcal{I}M$. Going back to the proof of Theorem 8.5 we can choose $a_{ij} \in \mathcal{I}$. In this case, the non-dominant coefficients of $P(X) = \det(X I_{n \times n} - (a_{ij}))$ are in $\mathcal{I}$.   □

8.2   Finite extensions, integral extensions

Definition 8.8 *Let A be a subring of B. Then B is algebraic (resp. integral) over A if x is algebraic (resp. integral) over A, for all $x \in B$. We say in this case that $A \subset B$ is an algebraic (resp. integral) extension.*
    If $f : A \to B$ is a ring homomorphism such that B is algebraic (resp. integral) over $f(A)$, we say that f is algebraic (resp. integral); and sometimes that B is algebraic (resp. integral) over A.

Definition 8.9 *A ring extension $A \subset B$ is finite if B is a finitely generated A-module.*

Proposition 8.10 *Let $A \subset B$ be an integral (resp. finite) ring extension.*

(i) *If $\mathcal{J}$ is an ideal of B, then $B/\mathcal{J}$ is integral (resp. finite) over $A/(\mathcal{J} \cap A)$.*

(ii) *If S is a multiplicatively closed part of A, then $S^{-1}B$ is integral (resp. finite) over $S^{-1}A$.*

Proof (i) is obvious. For (ii), consider $x/s \in S^{-1}B$. Since x is integral over A, there is a relation $x^n + a_1 x^{n-1} + \ldots + a_{n-1} x + a_n = 0$, with coefficients $a_i \in A$. This induces another relation

$$(x/s)^n + (a_1/s)(x/s)^{n-1} + \ldots + (a_{n-1}/s^{n-1})(x/s) + a_n/s^n = 0,$$

which proves that x/s is integral over $S^{-1}A$. We have already seen that if $x_1, \ldots, x_n$ generate B, as an A-module, then $x_1/1, \ldots, x_n/1$ generate $S^{-1}B$ as an $S^{-1}A$-module.   □

The following important result is an immediate consequence of Theorem 8.5.

Proposition 8.11 *A finite ring extension is integral.*

Proposition 8.12 *If $A \subset B$ and $B \subset C$ are finite ring extensions, then $A \subset C$ is a finite ring extension.*

Proof Let $(b_1, \ldots, b_r)$ (resp. $(c_1, \ldots, c_m)$) be a system of generators of B (resp. C) as an A-module (resp. B-module). We claim that $(b_j c_i)_{i,j}$ is a system of generators of C as an A-module. Indeed, consider $z \in C$. There are elements $y_i \in B$ such that $z = \sum y_i c_i$ and $x_{ij} \in A$ such that $y_i = \sum x_{ij} b_j$. This shows $z = \sum x_{ij} b_j c_i$.   □

Corollary 8.13 *If $A \subset B$ is a ring extension and $x_1, \ldots, x_n \in B$ are integral over A, then $A \subset A[x_1, \ldots, x_n]$ is a finite extension.*

Proof We have already proved this result for $n = 1$. Assume it is true for $n - 1$. Since x_n is integral over A, it is integral over $A[x_1, \ldots, x_{n-1}]$. The extension $A[x_1, \ldots, x_{n-1}] \subset A[x_1, \ldots, x_n]$ is therefore finite, and we conclude with Proposition 8.12.   □

Corollary 8.14 *Let $A \subset B$ be a ring extension. The set of all elements of B integral over A is a subring of B (the integral closure of A in B).*

Proof If $x, y \in B$ are integral over A, then $A[x, y]$ is finite over A, hence $x - y$ and xy are integral over A.   □

Corollary 8.15 *If $A \subset B$ and $B \subset C$ are integral ring extensions, then $A \subset C$ is an integral ring extension.*

Proof Consider $z \in C$ and $z^n + b_1 z^{n-1} + \ldots + b^n = 0$ a relation of integral dependence of z over B. Clearly z is integral over the subring $B' = A[b_1, b_2, \ldots, b_n]$ of B. By Corollary 8.13, B' is finite over A. Since $B'[z]$ is finite over B', it is finite over A by Proposition 8.12 and z is integral over A by 8.5. $\qquad\square$

Exercise 8.16 If A' is the integral closure of A in B and S a multiplicatively closed part of A, show that $S^{-1}A'$ is the integral closure of $S^{-1}A$ in $S^{-1}B$.

Definition 8.17 (i) *If the integral closure of A in B is A itself, we say that A is integrally closed in B.*

(ii) *A domain A is integrally closed if it is integrally closed in its fractions field $K(A)$.*

Exercises 8.18

1. Show that the integral closure of A in B is integrally closed in B.

2. Show that a UFD is integrally closed.

Proposition 8.19 *If A is an integrally closed domain, $A[X]$ is integrally closed.*

Proof Let K be the fraction field of A. Since $K[X]$ is integrally closed, it is sufficient to show that if $P \in K[X]$ is integral over $A[X]$, then $P \in A[X]$. Consider the relation of integral dependence $P^n + a_1 P^{n-1} + \cdots + a_n = 0$, with $a_i \in A[X]$. If m is a positive integer and $Q = P - X^m$, this relation induces a relation

$$Q^n + b_1 Q^{n-1} + \cdots + b_n = 0, \quad \text{with} \quad b_n = a_n - a_{n-1}X^m - \cdots - a_1 X^{m(n-1)} - X^{nm}.$$

Fixing m large enough, the polynomials $-Q$ and $-b_n$ are monic.

We admit here that K is contained in an algebraically closed field L. In $L[X]$, there is a decomposition

$$-Q = \prod_1^m (X - d_i) = X^m - (\sum_1^m d_i)X^{m-1} + (\sum_{i<j} d_i d_j)X^{m-2} + \cdots + (-1)^m d_1 \ldots d_m.$$

Now it is clear that a root of Q is a root of b_n. This shows that the elements $d_i \in L$ are integral over A. Consequently the coefficients of Q are integral over A. Since these coefficients are in K, they are in A and $Q \in A[X]$. This shows $P \in A[X]$ and we are done. $\qquad\square$

Proposition 8.20

(i) *If A is integrally closed in B and $S \subset A$ is a multiplicatively closed part, $S^{-1}A$ is integrally closed in $S^{-1}B$.*

(ii) *If A is an integrally closed domain and $S \subset A$ is a multiplicatively closed part, $S^{-1}A$ is an integrally closed domain.*

(iii) *Let $A \subset B$ be a ring extension. If for all maximal ideals $\mathcal{M}$ of A, the local ring $A_{\mathcal{M}}$ is integrally closed in $S^{-1}B$, where $S = A - \mathcal{M}$, then A is integrally closed in B.*

(iv) *Let A be a domain. If for all maximal ideals $\mathcal{M}$ of A, the local ring $A_{\mathcal{M}}$ is integrally closed, then A is integrally closed.*

Proof (i) is a special case of Exercise 8.16 and (ii) a special case of (i).

For (iii), consider $x \in B$ an element integral over A. Let $\mathcal{M}$ be a maximal ideal of A. There exist $a \in A$ and $s \in A - \mathcal{M}$ such that $x/1 = a/s \in A_{\mathcal{M}}$. As a consequence, there exists $t \in A - \mathcal{M}$ such that $tx \in A$. This shows that the conductor $(A : x)$ of x in A is not contained in any maximal ideal of A, hence $A : x = A$ and $x \in A$. Finally, (iv) is a special case of (iii).   □

8.3   Going-up and going-down theorems

We conclude this chapter with two famous theorems. The going-up is an immediate consequence of Theorem 8.21 below. The proof is straightforward. The going-down is much more subtle. The proof given here is short. It does not clearly show how this result is related to Galois theory. We shall come back to this aspect in chapter 14, section 5.

Theorem 8.21 *If $A \subset B$ is an integral extension of domains, A is a field if and only if B is a field.*

Proof Assume first that A is a field. Consider $x \in B$ and the relation of integral dependence $x^n + a_1 x^{n-1} + \ldots + a_{n-1}x + a_n = 0$ of x over A. Simplifying, if necessary, by a power of x, we can assume $a_n \neq 0$. This shows $x(x^{n-1} + a_1 x^{n-2} + \ldots + a_{n-1})a_n^{-1} = -1$, hence x is invertible.

Conversely, assume B is a field. Let $a \in A$ be a non-zero element. Since $a^{-1} \in B$, there exists a relation $a^{-n} + a_1 a^{1-n} + \ldots + a_{n-1}a^{-1} + a_n = 0$, with $a_i \in A$. Multiplying by a^{n-1}, we find $a^{-1} = -(a_1 + a_2 a + \ldots + a_{n-1}a^{n-2} + a_n a^{n-1})$, which shows $a^{-1} \in A$.   □

Corollary 8.22 *Let $K \subset L$ be a field extension and $x \in L$. Then x is algebraic over K if and only if $K[x]$ is a field.*

Proof If x is algebraic, hence integral, over K, then $K \subset K[x]$ is an integral extension of domains. By Theorem 8.21, $K[x]$ is a field.

If x is not algebraic, the isomorphism $K[X] \simeq K[x]$ shows that $K[x]$ is not a field. □

Corollary 8.23 *Let $A \subset B$ be an integral ring extension. If $\mathcal{P}$ is a prime ideal of B, then $\mathcal{P}$ is maximal if and only if $\mathcal{P} \cap A$ is a maximal ideal of A.*

Proof Since $A/(\mathcal{P} \cap A) \subset B/\mathcal{P}$ is an integral ring extension of domains, we are done by Theorem 8.21. □

Exercise 8.24 Let A be a domain. Show that if a finitely generated A-algebra $A[x_1, ..., x_r]$ is a field algebraic over A, there exists $s \in A$ such that A_s is a field.

Be sure to solve the next exercise. You will have proved Hilbert's Nullstellensatz. We come back to this result in chapter 10, where we give a different proof.

Exercise 8.25 Let K be a field. Show, by induction on n, that if a finitely generated K-algebra $K[x_1, \ldots, x_n]$ is a field, it is an algebraic extension of K.

Hint: by Corollary 8.22, this is true for $n = 1$. By the induction hypothesis and by Exercise 8.24, there exists $s \in K[x_1]$ such that $K[x_1]_s$ is a field. Use this fact to show that $K[x_1]$ is a field and conclude.

Theorem 8.26 *Let $A \subset B$ be an integral ring extension.*

(i) *If $\mathcal{P}$ is a prime ideal of A, there is a prime ideal $\mathcal{P}'$ of B such that $\mathcal{P}' \cap A = \mathcal{P}$.*

(ii) *If $\mathcal{P}' \subset \mathcal{P}''$ are prime ideals of B such that $\mathcal{P}' \cap A = \mathcal{P}'' \cap A$, then $\mathcal{P}' = \mathcal{P}''$.*

Proof Assume first that $\mathcal{P}$ is the only maximal ideal of A. If $\mathcal{M}$ is a maximal ideal of B, we have $\mathcal{M} \cap A = \mathcal{P}$, by the preceding corollary, and (i) is proved in this case.

In general, consider $S = A \setminus \mathcal{P}$ and the integral ring extension

$$A_{\mathcal{P}} \subset S^{-1}B.$$

Since $A_{\mathcal{P}}$ is local with maximal ideal $\mathcal{P}A_{\mathcal{P}}$, there exists a maximal ideal $\mathcal{M}$ of $S^{-1}B$ such that $\mathcal{M} \cap A_{\mathcal{P}} = \mathcal{P}A_{\mathcal{P}}$. Putting $\mathcal{P}' = \mathcal{M} \cap B$, we have

$$\mathcal{P}' \cap A = \mathcal{M} \cap B \cap A = \mathcal{M} \cap S^{-1}A \cap A = \mathcal{P}S^{-1}A \cap A = \mathcal{P}.$$

For (ii), consider $\mathcal{P} = \mathcal{P}' \cap A = \mathcal{P}'' \cap A$ and $S = A \setminus \mathcal{P}$. We have

$$\mathcal{P}S^{-1}A = \mathcal{P}'S^{-1}B \cap S^{-1}A = \mathcal{P}''S^{-1}B \cap S^{-1}A.$$

Since $\mathcal{P}S^{-1}A$ is a maximal ideal of $S^{-1}A$, the prime ideals $\mathcal{P}'S^{-1}B$ and $\mathcal{P}''S^{-1}B$ of $S^{-1}B$ are maximal by Corollary 8.23. The inclusion $\mathcal{P}'S^{-1}B \subset \mathcal{P}''S^{-1}B$ proves $\mathcal{P}'S^{-1}B = \mathcal{P}''S^{-1}B$, hence $\mathcal{P}' = \mathcal{P}''$. □

Corollary 8.27 *(Going-up theorem) If $\mathcal{P}_0 \subset \mathcal{P}_1$ is an increasing sequence of prime ideals of A and if $\mathcal{P}'_0$ is a prime ideal of B such that $\mathcal{P}'_0 \cap A = \mathcal{P}_0$, there exists a prime ideal $\mathcal{P}'_1$ of B such that*

$$\mathcal{P}'_0 \subset \mathcal{P}'_1 \text{ and } \mathcal{P}'_1 \cap A = \mathcal{P}_1.$$

Proof Since $A/\mathcal{P}_0 \subset B/\mathcal{P}'_0$ is an integral ring extension, there exists a prime ideal $\mathcal{N}$ of $B/\mathcal{P}'_0$ such that $\mathcal{N} \cap (A/\mathcal{P}_0) = \mathcal{P}_1/\mathcal{P}_0$. If $\mathcal{P}'_1$ is the prime ideal of B such that $\mathcal{P}'_1/\mathcal{P}'_0 = \mathcal{N}$, it obviously satisfies our assertion. □

Exercise 8.28 Show that if a domain A is integral over a principal ideal ring R, all non-zero prime ideals of A are maximal.

Theorem 8.29 *(Going-down theorem) Let A be an integrally closed ring and $A \subset B$ an integral extension of domains. If $\mathcal{P}_1 \subset \mathcal{P}_0$ is a decreasing sequence of prime ideals of A and if $\mathcal{P}'_0$ is a prime ideal of B such that $\mathcal{P}'_0 \cap A = \mathcal{P}_0$, there exists a prime ideal $\mathcal{P}'_1$ of B such that*

$$\mathcal{P}'_1 \subset \mathcal{P}'_0 \text{ and } \mathcal{P}'_1 \cap A = \mathcal{P}_1.$$

The proof of this last theorem is delicate and important. It relies on the two following lemmas.

Lemma 8.30 *Let $A \subset B$ be a ring extension and $\mathcal{P}$ a prime ideal of A. Then there exists a prime ideal $\mathcal{P}'$ of B such that $\mathcal{P}' \cap A = \mathcal{P}$ if and only if $\mathcal{P}B \cap A = \mathcal{P}$.*

Proof Assume $\mathcal{P}' \cap A = \mathcal{P}$. Then $\mathcal{P} \subset \mathcal{P}B \cap A \subset \mathcal{P}' \cap A = \mathcal{P}$ shows $\mathcal{P}B \cap A = \mathcal{P}$.

Conversely, assume $\mathcal{P}B \cap A = \mathcal{P}$. If $\mathcal{P}$ is a maximal ideal of A, letting $\mathcal{P}'$ be a maximal ideal of B containing $\mathcal{P}B$, we have $\mathcal{P}' \cap A = \mathcal{P}$.

In general, consider the multiplicatively closed part $S = A \setminus \mathcal{P}$. We have $\mathcal{P}S^{-1}B \cap B = \mathcal{P}B$ by Corollary 7.45. Let $\mathcal{N}$ be a maximal ideal of $S^{-1}B$ containing $\mathcal{P}S^{-1}B$. If $\mathcal{P}' = \mathcal{N} \cap B$, we have $\mathcal{P}B \subset \mathcal{P}'$ on the one hand and $\mathcal{P}' \cap S = \emptyset$ on the other. This implies $\mathcal{P} \subset (\mathcal{P}' \cap A)$ and $(\mathcal{P}' \cap A) \cap S = \emptyset$, hence $\mathcal{P}' \cap A = \mathcal{P}$. □

Lemma 8.31 *Let A be an integrally closed ring, $A \subset B$ an integral extension of domains and $x \in B$. If $P \in K(A)[X]$ is the minimal polynomial of x over the fractions field $K(A)$ of A, then:*

(i) *the coefficients of P are in A;*

(ii) *if furthermore $\mathcal{P}$ is a prime ideal of A such that $x \in \mathcal{P}B$, the non-dominant coefficients of P are in $\mathcal{P}$.*

Proof We admit that every field is contained in an algebraically closed field.

If L is an algebraically closed field containing the fraction field $K(B)$, there are elements $x_1, x_2, ..., x_n \in L$ such that $P = \prod_1^n (X - x_i)$. This shows

$$(*) \quad P(X) = X^n - (\sum_1^n x_i)X^{n-1} + (\sum_{i<j} x_ix_j)X^{n-2} + ... + (-1)^n x_1x_2...x_n.$$

Consider $x^r + b_1 x^{r-1} + ... + b_r = 0$, a relation of integral dependence of x over A. The polynomial $Q = X^r + b_1 X^{r-1} + ... + b_r$, whose coefficients are in $A \subset K(A)$, is a multiple of P in $K(A)[X]$. This proves $Q(x_i) = 0$, hence that x_i is integral over A, for all i. Going back to the description of P, its coefficients are also integral over A. Since A is integrally closed and these coefficients are in $K(A)$, we have proved (i).

Assume furthermore $x \in \mathcal{P}B$. We know from Proposition 8.7 that x satisfies an integral dependence relation $x^r + b_1 x^{r-1} + ... + b_r = 0$, with $b_i \in \mathcal{P}$. As before, this shows, for all i,

$$x_i^r + b_1 x_i^{r-1} + ... + b_r = 0.$$

If $B' = A[x_1, ..., x_n]$, it proves $x_i^r \in \mathcal{P}B'$, for all i. Since B' is integral over A, there exists a prime ideal $\mathcal{P}'$ of B' such that $\mathcal{P}' \cap A = \mathcal{P}$. We have $x_i^r \in \mathcal{P}'$ hence $x_i \in \mathcal{P}'$, for all i. Using $(*)$ again, we see that the non-dominant coefficients of P are in $\mathcal{P}' \cap A = \mathcal{P}$, and (ii) is proved. $\square$

Proof of Theorem 8.29: To begin with, note that it is sufficient to prove that there exists a prime ideal $\mathcal{N}$ of $B_{\mathcal{P}'_0}$ such that $\mathcal{N} \cap A = \mathcal{P}_1$. Indeed, $\mathcal{P}'_1 = \mathcal{N} \cap B$ will satisfy the theorem. Using Lemma 8.30, it is therefore enough to show $\mathcal{P}_1 B_{\mathcal{P}'_0} \cap A = \mathcal{P}_1$.

Consider an element $x/s \in \mathcal{P}_1 B_{\mathcal{P}'_0} \cap A$, with $x \in \mathcal{P}_1 B$ and $s \in (B \setminus \mathcal{P}'_0)$. We use first that $a = x/s \in A \subset K(A)$. This implies $K(A)[x] = K(A)[sa] = K(A)[s]$. Consequently, the minimal polynomials of x and s over $K(A)$ have the same degree, by Proposition 8.4.

Let $P = X^n + b_1 X^{n-1} + ... + b_n$ be the minimal polynomial of x over $K(A)$. By Lemma 8.31, we have $b_i \in \mathcal{P}_1$, for all i. Putting $Q(X) = X^n + (b_1/a)X^{n-1} +$

$\dots + (b_n/a^n)$, it is clear that $Q(s) = 0$, hence Q is the minimal polynomial of s. Using Lemma 8.31 once more, we see that $(b_i/a^i) \in A$, for all i. Assume $a \notin \mathcal{P}_1$. Since $b_i = a^i(b_i/a^i) \in \mathcal{P}_1$, this implies $(b_i/a^i) \in \mathcal{P}_1$. With $Q(s) = 0$, this shows $s^n \in \mathcal{P}_1 B \subset \mathcal{P}'_0$. This is a contradiction and Theorem 8.29 is proved.

8.4   Exercises

1. Let A be a ring and M an A-module generated by n elements. Show that for any endomorphism u of M there exists a monic polynomial $P \in A[X]$ satisfying $P(u) = 0$ and such that $\deg P \le n$.

2. Let $A \subset B$ be an extension of domains. Assume that the ideal $A : B = \{a \in A,\ aB \subset A\}$ of A is finitely generated and show that B is integral over A.

3. Let A be a domain and $a \in A$ an element such that the ring A/aA is reduced. Show that if A_a is integrally closed, then so is A.

   Show that the ring $R = \mathbb{C}[X, Y, Z]/(XZ - Y(Y+1))$ is integrally closed.

4. Let $A \subset B$ be a ring extension. Assume that for each minimal prime ideal $\mathcal{P}$ of B, the ring extension $A/(\mathcal{P} \cap A) \subset B/\mathcal{P}$ is integral. Show that B is integral over A. Hint: for $x \in B$ consider the elements $P(x)$ for all monic polynomials $P \in A[X]$ and show that they form a multiplicatively stable part of B that contains $\{0\}$.

5. Let $A \subset B$ be a finite extension of rings. If $\mathcal{P}$ is a prime ideal of A, put $S = A \setminus \mathcal{P}$ and show that the ring $S^{-1}(B/\mathcal{P}B)$ is Artinian. Define $e(\mathcal{P})$ to be the length $\mathrm{l}(S^{-1}(B/\mathcal{P}B))$ and show that the function e defined on $\mathrm{Spec}(A)$ is semi-continuous.

   Assume furthermore that A is a domain and show that if the function e is constant, then B is locally free as an A-module.

6. Let A be an integrally closed domain. If $A[X] \subset B$ is an integral extension of domains, show that for any maximal ideal $\mathcal{M}$ of B, there exists a prime ideal $\mathcal{P}$ of B strictly contained in $\mathcal{M}$ and such that $\mathcal{P} \cap A = \mathcal{M} \cap A$.

7. Let A be a domain. Assume that for each non-zero element $a \in K(A)$ such that $a \notin A$, then $a^{-1} \in A$. Show that A is integrally closed.

8. Let A be an integrally closed domain and $B = A[x]$ a domain integral over A. Show that B is a free A-module.

9

Algebraic extensions of fields

In complex geometry we only need some basic results of Galois theory. Characteristic zero really makes life easier. The content of this chapter can be found in all algebra books, but in order to keep this book self-contained we present here precisely the results that we shall need when we start studying complex geometry.

An algebraic extension of fields is an integral extension of rings. Hence several of the results presented in chapter 8 will be useful now. For the convenience of the reader, we recall some of these results in the form needed here.

9.1 Finite extensions

Proposition 9.1 *Let $K \subset L$ be a field extension and $x \in L$. The following conditions are equivalent:*

(i) *the element x is algebraic over K;*

(ii) *the K-vector space $K[x]$ has finite rank;*

(iii) *the K-algebra $K[x]$ is a field.*

The equivalence of (i) and (ii) is a special case of Theorem 8.5; that of (i) and (iii) is Corollary 8.22.

Corollary 9.2 *If $K \subset L$ is a finite field extension, L is algebraic over K.*

Note that this corollary is a special case of Proposition 8.11.

Proposition 9.3 *If $K \subset L$ is a field extension and if $x_1, \ldots, x_n \in L$ are elements algebraic over K, then $K[x_1, \ldots, x_n]$ is a field.*

Proof Since $K[x_1, \ldots, x_n]$ is a domain and is integral over K, this is a special case of Theorem 8.21. $\qquad\square$

Definition 9.4 *Let $K \subset L$ be a finite field extension. The degree of this extension, $\deg_K(L)$, also denoted by $[L : K]$, is the rank of the K-vector space L.*

If there exists $x \in L$ such that $L = K[x]$, we define $\deg_K(x) = [L : K]$.

Proposition 9.5 *If P is the minimal polynomial of x over K, then*

$$K[X]/(P) \simeq K[x] \quad and \quad \deg_K(x) = \deg(P).$$

This is Proposition 8.4.

Proposition 9.6 *Let $K \subset L \subset F$ be field extensions.*

(i) *If L is algebraic over K and F is algebraic over L, then F is algebraic over K.*

(ii) *If both extensions are finite, F is finite over K and*

$$[F : K] = [F : L][L : K].$$

Proof   (i) is a special case of Corollary 8.15.

For (ii), note that if (e_i) is a basis of F over L and (f_j) a basis of L over K, then it is straightforward to verify that $(e_i f_j)$ is a basis of F over K.   □

Corollary 9.7 *If $K \subset L$ is a finite field extension and $x \in L$, then $\deg_K(x)$ divides $[L : K]$.*

Proof   This is clear.   □

We recall that a field Ω is algebraically closed if each non-constant polynomial with coefficients in Ω has a root in Ω, or equivalently if all irreducible polynomials in $\Omega[X]$ have degree 1.

Definition 9.8 *If $K \subset \Omega$ is an algebraic field extension such that Ω is algebraically closed, Ω is called an algebraic closure of K.*

Definition 9.9 *Let Ω be an algebraic closure of K and $P \in K[X]$ a non-zero polynomial. If $P = u \prod_{i=1}^{n}(X - x_i)$ (with $u \in K$) is the decomposition of P as a product of irreducible polynomials in $\Omega[X]$, then $K[x_1, ..., x_n]$ is the field of decomposition of P in Ω.*

Clearly, the field of decomposition of P in Ω is the smallest subfield of Ω containing K and in which P decomposes into degree one factors.

We shall assume the next theorem without proof. An excellent proof can be found in Bourbaki, *Algèbre*, chap. 5, p. 91, but the result is all we need.

Theorem 9.10 *Every field has an algebraic closure.*

Definition 9.11 *Let $K \subset L$ and $K \subset F$ be field extensions. A field homomorphism $f : L \to F$ such that $f(a) = a$ for all $a \in K$ is called a K-isomorphism.*

Note that a K-isomorphism is always injective, but not necessarily surjective. Yes this is unpleasant, but that's life!

Theorem 9.12 *Let Ω be an algebraic closure of K and $K \subset L \subset L'$ be algebraic extensions. If $u : L \to \Omega$ is a K-isomorphism, there exists a K-isomorphism $u' : L' \to \Omega$ such that $u'(x) = u(x)$ for all $x \in L$. We say that u' extends u.*

This result is easy to prove when the extension $L \subset L'$ is finite. If not, things get more complicated. Since we're in the mood, let's assume this theorem as well (for a proof, I recommend once more Bourbaki, *Algèbre*, chap. 5, p. 90).

Definition 9.13 *Let Ω be an algebraic closure of K. If $K \subset L \subset \Omega$ and $K \subset F \subset \Omega$ are extensions, they are conjugate if there exists a surjective K-isomorphism $u : L \to F$.*

Note that since u is an isomorphism of fields, there is an inverse isomorphism $u^{-1} : F \to L$. Furthermore $u(a) = a$ for all $a \in K$ implies obviously $u^{-1}(a) = a$ for all $a \in K$ and u^{-1} is a K-isomorphism.

Note also that by Theorem 9.12, u extends to a K-isomorphism of Ω to itself.

Definition 9.14 *Let x and y be elements of Ω. We say that x and y are conjugate over K, if there exists a K-isomorphism $u : K[x] \to K[y]$ such that $u(x) = y$.*

Note now that if x and y are conjugate, then $u(K[x]) = K[y]$ and that $u^{-1} : K[y] \to K[x]$ is a K-isomorphism such that $u^{-1}(y) = x$.

Proposition 9.15 *Two elements x and y are conjugate over K if and only if they have the same minimal polynomial over K.*

Proof Let $P \in K[X]$ and $Q \in K[X]$ be the minimal polynomials of x and y over K.

If x and y are conjugate by the K-isomorphism u, then

$$0 = u(0) = u(P(x)) = P(u(x)) = P(y).$$

Hence Q divides P. But P divides Q for the same reason, so $P = Q$ since they are both monic.

Conversely, if $P = Q$, consider the natural K-isomorphisms

$$f : K[X]/(P) \simeq K[x] \quad \text{and} \quad g : K[X]/(P) \simeq K[y].$$

Clearly, the K-isomorphism $g \circ f^{-1}$ satisfies $g \circ f^{-1}(x) = y$. □

9.2 K-isomorphisms in characteristic zero

Proposition 9.16 *Let K be a field of characteristic zero and $P \in K[X]$ an irreducible polynomial. If $K \subset L$ is a field extension such that P decomposes into a product of degree one polynomials in $L[X]$, then P has no multiple root in L.*

Proof Let $x \in L$ be a root of P. Then P is necessarily the minimial polynomial (up to a constant) of x. Since char$(K) = 0$, the derivative $P' \in K[X] \subset L[X]$ of P is non-zero. Consequently, P' is not a multiple of P and x is not a root of P'.   □

Corollary 9.17 *Let Ω be an algebraic closure of a field K of characteristic zero. If $P \in K[X]$ is irreducible, then P admits $\deg(P)$ distinct roots in Ω.*

Proof This is a special case of Proposition 9.16.   □

Proposition 9.18 *Let K be a field of characteristic zero and Ω an algebraic closure of K. If $K \subset L$ is an algebraic field extension and $x \in L$, then the number of distinct K-isomorphisms $K[x] \rightarrow \Omega$ is precisely $[K[x] : K]$.*

Proof Put $n = [K[x] : K]$. The minimal polynomial of x over K has n distinct roots $y_1, ..., y_n \in \Omega$ by Corollary 9.17. By Proposition 9.15, there are K-isomorphisms $u_i : K[x] \rightarrow \Omega$ such that $u_i(x) = y_i$, for $i = 1, ..., n$, and no other.   □

Lemma 9.19 *Let $K \subset L$ be a finite extension of fields of characteristic zero. There exist $[L : K]$ distinct K-isomorphisms $L \rightarrow \Omega$.*

Proof We use induction on $n = [L : K]$. The result is obvious for $n = 0$. Consider $x \in L$, with $x \notin K$. Put $F = K[x]$, $m = [F : K]$ and $d = [L : F]$. Then $n = md$, with $d < n$.

From Proposition 9.18, there exist m distinct K-isomorphisms $u_i : F \rightarrow \Omega$, with $i = 1, ..., m$. By Theorem 9.12, for all i the K-isomorphism $u_i : F \rightarrow \Omega$ extends to a K-isomorphism $v_i : L \rightarrow \Omega$. Put $F_i = u_i(F)$ and $L_i = v_i(L)$ and note that $[L_i : F_i] = d$.

By the induction hypothesis, there exist d distinct F_i-isomorphisms $s_{ij} : L_i \rightarrow \Omega$, with $j = 1, ..., d$.

Define $t_{ij} = s_{ij} \circ v_i$, for $1 \leq i \leq m$ and $1 \leq j \leq d$. We claim that these K-isomorphisms $L \rightarrow \Omega$ are pairwise distinct.

Since $t_{ij}/F = u_i$, we have $t_{ij} \neq t_{i'j'}$ if $i \neq i'$, for any (j, j'). But $j \neq j'$ implies $s_{ij} \neq s_{ij'}$, hence $t_{ij} \neq t_{ij'}$ and we are done.   □

Theorem 9.20 *If $K \subset L$ is a finite extension of fields of characteristic zero, there exists $x \in L$ such that $L = K[x]$.*

Proof Put $n = [L : K]$. It is sufficient to find $x \in L$ such that $[K[x] : K] \geq n$; in other words, to find $x \in L$ having n distinct conjugates.

Let $u_i : L \to \Omega$, with $i = 1, ..., n$, be distinct K-isomorphisms. We are looking for an $x \in L$ such that $i \neq j$ implies $u_i(x) \neq u_j(x)$. Consider, for all (i, j), with $i \neq j$, the K-vector space $L_{ij} = \ker(u_i - u_j)$. Since $u_i \neq u_j$, we have $L_{ij} \neq L$. We want to show $L \neq \cup L_{ij}$. This is a consequence of the following general lemma. $\square$

Lemma 9.21 *Let k be an infinite field and E a finite rank k-vector space. If E_i, with $i = 1, ..., n$, are strict subvector spaces of E, then*

$$E \neq \cup_i E_i.$$

Proof We use induction on n. The result is obvious for $n = 1$. Assume $n > 1$.

By the induction hypothesis, there exists $x \in E$ such that $x \notin \cup_1^{n-1} E_i$. If $x \notin E_n$, we are done. If $x \in E_n$ consider $y \notin E_n$. Note that $ax + y \notin E_n$ for all $a \in k$. If the lemma is not true, then for any $a \in k$ there exists $i \leq n - 1$ such that $ax + y \in E_i$. Since k is infinite, there exist i and distinct elements $a, a' \in k$ such that $ax + y \in E_i$ and $a'x + y \in E_i$. This proves $(a - a')x \in E_i$, hence $x \in E_i$, a contradiction. $\square$

Corollary 9.22 *Let $K \subset L$ be a finite extension of fields of characteristic zero. The number of distinct K-isomorphisms $L \to \Omega$ is $[L : K]$.*

Proof This is an immediate consequence of Proposition 9.18 and Theorem 9.20. $\square$

9.3 Normal extensions

In this section K is a given field of characteristic zero.

Definition 9.23 *An algebraic extension $K \subset L$ is called normal if for any $x \in L$, the minimal polynomial $P \in K[X]$ of x over K decomposes into a product of degree one polynomials in $L[X]$.*

Note that $K \subset L$ is normal if for all $x \in L$, the field L contains a decomposition field of the minimal polynomial of x over K.

Examples 9.24

1. If $[L : K] = 2$, then $K \subset L$ is normal.

2. An algebraic closure Ω of K is a normal extension of K.

Proposition 9.25 *Assume $K \subset F \subset L$ are algebraic extensions. If $K \subset L$ is normal, then $F \subset L$ is normal.*

The proof is left to the reader.

Proposition 9.26 *Let $K \subset L$ be an algebraic extension and Ω an algebraic closure of L. Then $K \subset L$ is normal if and only if $u(L) \subset L$ for any K-isomorphism $u : L \to \Omega$.*

Proof Consider $x \in L$ and $P \in K[X]$ its minimal polynomial.

Assume that the extension $K \subset L$ is normal. Then $P = \prod(X - x_i)$, with $x_i \in L$. If $u : L \to \Omega$ is a K-isomorphism, then $u(x) = x_i$ for some i, by Proposition 9.15. This shows $u(L) \subset L$.

Conversely, consider $y \in \Omega$ a root of P. By Proposition 9.15, there exists a K-isomorphism $v : K[x] \to \Omega$ such that $v(x) = y$. By Theorem 9.12, there exists a K-isomorphism $u : L \to \Omega$ such that $u(x) = y$. Since $u(L) \subset L$, this proves $y \in L$. $\qquad\square$

Corollary 9.27 *If $P \in K[X]$, its decomposition field in an algebraic closure Ω of K is a normal extension of K.*

Proof Let $x_1, x_2, ..., x_n \in \Omega$ be the roots of P. Then $L = K[x_1, ..., x_n]$ is the decomposition field of P. If $u : L \to \Omega$ is a K-isomorphism, for all i there exists j such that $u(x_i) = x_j$. This proves $u(L) \subset L$ and we can apply Proposition 9.26. $\qquad\square$

Corollary 9.28 *Let Ω be an algebraic closure of K and $x \in \Omega$. If $P \in K[X]$ is the minimal polynomial of x over K, then the field of decomposition of P in Ω is the smallest normal extension of K contained in Ω and containing x.*

Proof A normal extension of K containing x must contain the field of decomposition of P. Since this field is a normal extension of K, we are done. $\qquad\square$

Definition 9.29 *If $K \subset L$ is an algebraic extension, a K-automorphism g of L is a field automorphism of L such that $g(a) = a$ for all $a \in K$.*

It is clear that the K-automorphisms of L form a group.

Definition 9.30 *If L is a field and G a group of field automorphisms of L, we denote by L^G the subfield of L formed of all elements $a \in L$ such that $g(a) = a$ for all $g \in G$.*

Note that if G is the group of K-automorphisms of L, then obviously

$$K \subset L^G.$$

Proposition 9.31 *If $K \subset L$ is a normal extension and Ω an algebraic closure of L, a K-isomorphism $u : L \to \Omega$ is a K-automorphism of L.*

Proof We have $u(L) \subset L$ by Proposition 9.26. Since u is injective, we must show that it is surjective. If $[L : K]$ is finite, then $u(L) \simeq L$ implies $[u(L) : K] = [L : K]$, hence $[L : u(L)] = 1$ and $L = u(L)$. If not, consider $x \in L$. Let $F \subset L$ be the decomposition field of the minimal polynomial of x. Then $K \subset F$ is normal and $[F : K]$ is finite. This shows $u(F) = F$, hence $x \in u(L)$ and we are done. $\qquad\square$

Theorem 9.32 *Let $K \subset L$ be a finite extension and G the group of K-automorphisms of L. The following conditions are equivalent:*

(i) *the extension $K \subset L$ is normal;*

(ii) *the order $\mathrm{Ord}(G)$ of G is at least $[L : K]$;*

(iii) *one has $L^G = K$.*

Furthermore, if these conditions are satisfied, then $\mathrm{Ord}(G) = [L : K]$.

Proof (i) $\Rightarrow$ (ii). Let Ω be an algebraic closure of L. By Proposition 9.18 and Theorem 9.20, there are precisely $[L : K]$ distinct K-isomorphisms $L \to \Omega$. If $K \subset L$ is normal, each of them is a K-automorphism of L by Proposition 9.31, i.e. an element of G. Hence G has at least $[L : K]$ elements.

(ii) $\Rightarrow$ (iii). The elements $a \in L$ such that $g(a) = a$ for all $g \in G$ form a subfield F of L containing K. By Corollary 9.22, there are precisely $[L : F]$ distinct F-isomorphisms from L to Ω. Since each $g \in G$ is one of these, we have

$$[L : K] \leq \mathrm{Ord}(G) \leq [L : F] \leq [L : K].$$

This shows $[L : F] = [L : K] = \mathrm{Ord}(G)$, hence $K = F$.

(iii) $\Rightarrow$ (i). Let $g_1, ..., g_n$ be the elements of G. If $x \in L$, put $x_i = g_i(x)$. Consider the polynomial, with coefficients in L,

$$P = \sum_{i=o}^{n} a_i X^{n-i} = \prod_{i=1}^{n} (X - x_i).$$

Note that its coefficients

$$a_1 = -\sum_{i=1}^{n} x_i, \quad a_2 = \sum_{i \neq j} x_i x_j, \ldots$$

satisfy $g_k(a_i) = a_i$ for all k. Hence $a_i \in K$, for all i, by the hypothesis, and $P \in K[X]$. Since $P(x) = 0$, the minimal polynomial of x divides P. This shows that any element conjugate to x is among the x_i, hence in L. Consequently, $K \subset L$ is a normal extension. $\square$

Definition 9.33 *If $K \subset L$ is a normal extension, the group $G = \mathrm{Aut}_K(L)$ of K-automorphisms of L is the Galois group $\mathrm{Gal}(L/K)$ of this normal extension.*

We have therefore proved $\mathrm{Ord}(\mathrm{Gal}(L/K)) = [L : K]$ and $L^{\mathrm{Gal}(L/K)} = K$.

Theorem 9.34 *Let $K \subset L$ be a normal finite extension of fields of characteristic zero and $G = \mathrm{Gal}(L/K)$ its Galois group.*

(i) *If G' is a subgroup of G, then $L^{G'}$ is a subfield of L, containing K, such that*
$$\mathrm{Gal}(L/L^{G'}) = G'.$$

(ii) *If F is a field such that $K \subset F \subset L$, the Galois group $\mathrm{Gal}(L/F)$ is a subgroup of G such that*
$$L^{\mathrm{Gal}(L/F)} = F.$$

(iii) *A subgroup G' of G is normal if and only if $K \subset L^{G'}$ is a normal extension. In this case $\mathrm{Gal}(L^{G'}/K) \simeq G/G'$.*

Proof   (i) By the definition of $L^{G'}$, we see that G' is a subgroup of $\mathrm{Gal}(L/L^{G'})$. Since $\mathrm{Gal}(L/L^{G'})$ has order $[L : L^{G'}]$, it is sufficient to show $[L : L^{G'}] \leq \mathrm{Ord}(G')$.

There exists $x \in L$ such that $L = L^{G'}[x]$. Let $g_1, \ldots, g_r$ be the elements of G' and $x_i = g_i(x)$. The coefficients of the polynomial $P = \prod_{i=1}^{r}(X - x_i) \in L[X]$ are invariant for G'. Hence $P \in L^{G'}[X]$. Since x is a root of P, we get

$$[L : L^{G'}] = [L^{G'}[x] : L^{G'}] \leq \deg(P) = \mathrm{Ord}(G'),$$

and we are done.

(ii) is an immediate consequence of Theorem 9.32.

(iii) Assume G' is a normal subgroup of G. If $g \in G$ and $g' \in G'$, we have $g^{-1}g'g \in G'$, hence $g^{-1}g'g(x) = x$ for all $x \in L^{G'}$. This shows

$$g'(g(x)) = g(x) \text{ for all } g \in G, \ g' \in G' \text{ and } x \in L^{G'}.$$

Consequently,

$$g(x) \in L^{G'} \text{ for all } g \in G \text{ and } x \in L^{G'},$$

in other words,

$$g(L^{G'}) \subset L^{G'} \text{ for all } g \in G.$$

Hence $K \subset L^{G'}$ is a normal extension by Proposition 9.26.

Assume now that $L^{G'}$ is a normal extension of K. By Proposition 9.26, we have

$$g(L^{G'}) \subset L^{G'} \text{ for all } g \in G.$$

Hence for all $g \in G$ the restriction of g to $L^{G'}$ is in $\mathrm{Aut}_K(L^{G'}) = \mathrm{Gal}(L^{G'}/K)$. We have thus defined a group homomorphism $G \to \mathrm{Gal}(L^{G'}/K)$. Its kernel is $G' = \mathrm{Gal}(L/L^{G'})$, which is therefore a normal subgroup of G. We note furthermore that G/G' is isomorphic to a subgroup of $\mathrm{Gal}(L^{G'}/K)$. Now, we have

$$\mathrm{Ord}(G')\mathrm{Ord}(G/G') = \mathrm{Ord}(G) = [L : K] = [L : L^{G'}][L^{G'} : K].$$

Since $\mathrm{Ord}(G') = [L : L^{G'}]$, this proves

$$\mathrm{Ord}(G/G') = [L^{G'} : K] = \mathrm{Ord}(\mathrm{Gal}(L^{G'}/K)), \quad \text{hence} \quad G/G' \simeq \mathrm{Gal}(L^{G'}/K).$$

$\square$

9.4 Trace and norm

Once again, K is a given field of characteristic zero.

Definition 9.35 *Let $K \subset L$ be a finite normal extension of fields and G its Galois group. The trace and the norm, over K, of an element $x \in L$ are defined by*

$$\mathrm{Tr}_{L/K}(x) = \sum_{g \in G} g(x) \quad \text{and} \quad \mathrm{N}_{L/K}(x) = \prod_{g \in G} g(x).$$

It is clear that if $g \in G$, then $gG = G$. This shows $g(\mathrm{Tr}_{L/K}(x)) = \mathrm{Tr}_{L/K}(x)$ and $g(\mathrm{N}_{L/K}(x)) = \mathrm{N}_{L/K}(x)$ for all $g \in G$ and consequently that $\mathrm{Tr}_{L/K}(x) \in K$ and $\mathrm{N}_{L/K}(x) \in K$.

If $x \in K$, then $\mathrm{Tr}_{L/K}(x) = [L : K]x$ and $\mathrm{N}_{L/K}(x) = x^{[L:K]}$.

Proposition 9.36 **(i)** $\mathrm{Tr}_{L/K}(.)$ *is a linear form on the K-vector space L.*
(ii) $\mathrm{N}_{L/K}(xy) = \mathrm{N}_{L/K}(x)\mathrm{N}_{L/K}(y)$.

This is clear from the definition.

Theorem 9.37 *Let $K \subset L$ be a finite normal extension of fields, G its Galois group and $x \in L$.*

Consider L as a K-vector space and the endomorphism $u_x(y) = xy$ of L.

Let $P_{x,L}$ be the characterisitic polynomial of the endomorphism u_x and P be the minimal polynomial of x over K. If $d = \deg_K(x) = \deg(P)$ and $n = [L : K]$, we have

$$\prod_{g \in G} (X - g(x)) = P^{n/d} = P_{x,L}.$$

Proof Consider the subgroup $H = \mathrm{Gal}(L/K[x])$ of G. It is clear that an element $g \in G$ is in H if and only if $g(x) = x$. If we put $m = [L : K[x]]$, then $dm = n$.

We show $\prod_{g \in G}(X - g(x)) = P^m$. Consider $x_1, ..., x_d$ the distinct roots of P. Let $g_1, ..., g_d \in G$ be elements such that $g_i(x) = x_i$. If $g \in G$, there exists i such that $g(x) = x_i$. Furthermore,

$$g(x) = x_i \Leftrightarrow g \in g_i H.$$

This shows $G = \cup_1^d g_i H$. We note that this is a disjoint union and that $g_i H$ has m elements for all i. Consequently, we have

$$\prod_{g \in G} (X - g(x)) = \prod_{i=1}^d (\prod_{g \in g_i H} (X - g(x))) = \prod_{i=1}^d (X - x_i)^m = (\prod_{i=1}^d (X - x_i))^m = P^m.$$

Next, we prove $P_{x,L} = P^m$.

The K-vector space $K[x]$ has rank d and is stable for u_x. By the Cayley–Hamilton theorem, the characteristic polynomial of this endomorphism of $K[x]$ has degree $d = \deg(P)$ and x is a root of this polynomial. This proves that P is that polynomial.

Let $(e_1, ..., e_m)$ be a basis of L over $K[x]$. Clearly $K[x]e_i$ is stable for u_x in L. The natural $K[x]$-isomorphism $K[x] \simeq K[x]e_i$ shows that P is also the characteristic polynomial of the restriction of u_x to the stable vector space $K[x]e_i$. Considering then the decomposition of the vector space L, in u_x-stable factors, $L = \oplus_{i=1}^m K[x]e_i$, we find $P_{x,L} = P^m$. $\square$

Corollary 9.38 *Let $P = X^d - a_1 X^{d-1} + ... + (-1)^d a_d$ be the minimal polynomial of x over K. If $m = [L : K[x]]$, then*

$$\mathrm{Tr}_{L/K}(x) = ma_1 \quad and \quad \mathrm{N}_{L/K}(x) = a_d^m.$$

This is an immediate consequence of Theorem 9.37.

Theorem 9.39 *Let $K \subset L$ be a finite normal extension of fields of characteristic zero. Then*

$$(x, y) \to \mathrm{Tr}_{L/K}(xy)$$

is a symmetric non-degenerate bilinear form on the K-vector space L.

Proof The form is obviously symmetric and bilinear. We will need the following lemma.

Lemma 9.40 *Let $(z_1, ..., z_n)$ be a basis of the K-vector space L and $g_1, ..., g_n$ the elements of the Galois group $\mathrm{Gal}(L/K)$. There is a matrix equality*

$$(\mathrm{Tr}_{L/K}(z_i z_j))_{i,j} = (^{t}(g_i(z_j))_{i,j})((g_i(z_j))_{i,j}).$$

Proof of the lemma We have

$$\sum_l g_l(z_i) g_l(z_j) = \sum_l g_l(z_i z_j) = \mathrm{Tr}_{L/K}(z_i z_j).$$

$\square$

Proof of the theorem

Let $x \in L$ be such that $L = K[x]$ and $x_i = g_i(x)$, for $i = 1, ..., n$. The matrix $(g_i(x^j))_{i,j}$ associated to the basis $(1, x, ..., x^{n-1})$ of L is the Vandermonde matrix $(x_i^j)_{0 \le i,j \le (n-1)}$. Its determinant $\prod_{i<j}(x_i - x_j)$ is non-zero, since $x_i \ne x_j$ for $i \ne j$. $\square$

9.5 Roots of one and cyclic Galois groups

Once more K is a field of characteristic zero and Ω an algebraic closure of K.

Proposition 9.41 *The polynomial $X^n - 1 = 0$ has n distinct roots in Ω.*

This is clear since the unique root of $n X^{n-1} = 0$ is not a root of $X^n - 1 = 0$.

Our next assertion needs no comment.

Proposition 9.42 *The roots of $X^n - 1 = 0$ form a subgroup of the multiplicative group Ω^*.*

Definition 9.43 *A primitive dth root of 1, in Ω, is an element $a \in \Omega^*$ such that $a^d = 1$ and $a^m \ne 1$ for $0 < m < d$.*

Proposition 9.44 *A primitive dth root of 1, in Ω, generates the group of roots of $X^d = 1$.*

Proof A primitive dth root of 1, in Ω, generates a group of order d contained in the group of roots of $X^d = 1$, whose order is d.   □

Lemma 9.45 *If there exists one primitive dth root of 1 in Ω^*, there are precisely $\phi(d)$, where $\phi(d)$ is the number of integers relatively prime to d contained in $[0, d-1]$ for $1 < d$ and $\phi(1) = 1$.*

Proof If Ω^* contains a primitive dth root of 1, the roots of $X^d - 1 = 0$ form a cyclic group G, of order d. There is therefore an isomorphism $\pi : G \to \mathbb{Z}/d\mathbb{Z}$. Furthermore $a \in G$ is a primitive dth root of 1 if and only if $\pi(a)$ generates the additive group $\mathbb{Z}/d\mathbb{Z}$. Since a class generates $\mathbb{Z}/d\mathbb{Z}$ if and only if its elements are relatively prime to d, we are done.   □

Theorem 9.46 *A finite subgroup of Ω^* whose elements are roots of 1 is cyclic.*

Proof Let G be such a group and n its order. All elements in G are roots of $X^n - 1 = 0$. We claim that G contains a primitive nth root of 1.

If $a \in G$, there exists a positive integer d dividing n such that a is a primitive dth root of 1. Let $E_d \subset G$ be the set of all primitive dth roots of 1 contained in G. Clearly $G = \cup E_d$ for d dividing n. Since we have seen that E_d contains either 0 or $\phi(d)$ elements, we can conclude with the following lemma, whose proof is left to the reader.

Lemma 9.47 $\sum_{d|n} \phi(d) = n$.

In particular, we have proved the following statement.

Corollary 9.48 *The roots of $X^n - 1 = 0$ form a cyclic subgroup, of order n, of Ω^*.*

We can now describe, when K contains enough roots of 1, the normal extensions of K with cyclic Galois groups.

Theorem 9.49 *Assume K is a field such that $X^n - 1 \in K[X]$ decomposes in a product of degree 1 polynomials.*

(i) *If $a \in K$ and $x \in \Omega$ are such that $x^n = a \in K$, then the extension $K \subset K[x]$ is normal and its Galois group is cyclic. The order d of this group divides n and $x^d \in K$.*

(ii) *If $K \subset L$ is a finite normal extension with a cyclic Galois group and such that $d = [L : K]$ divides n, there exists $x \in L$ such that $x^d \in K$ and that $L = K[x]$.*

Proof (i) Let $\eta \in K$ be a primitive nth root of 1. Clearly $\eta^i x$ is a root of $X^n = a$. Thus this equation has n distinct roots in $K[x]$. Hence the minimal polynomial of x over K factorizes into degree 1 polynomials over $K[x]$ and $K \subset K[x]$ is normal. Let G be the Galois group of this extension and $g \in G$. There exists $i < n$ such that $g(x) = \eta^i x$. The map $g \to \eta^i = g(x)x^{-1}$ is obviously an injective group homomorphism from G into the group of nth roots of 1. Hence G is isomorphic to a subgroup of this group. By Theorem 9.46, there exists d, dividing n, such that G is isomorphic to the cyclic group of roots of $X^d - 1 = 0$. If ζ is a primitive dth root of 1, the conjugates of x are $\zeta^j x$ for $1 \le j \le d$. Hence

$$N_{K[x]/K}(x) = \prod_{j=1}^{d} \zeta^j x = (\prod_{j=1}^{d} \zeta^j) x^d = x^d.$$

Since $N_{K[x]/K}(x) \in K$, (i) is proved.

(ii) Let $\eta \in K$ be a primitive dth root of 1. Let τ be a generator of $\mathrm{Gal}(L/K)$. We claim that there exists a non-zero element $x \in L$ such that $\tau(x) = \eta x$.

Notice first that (ii) is an immediate consequence of the existence of such an x. Indeed, it shows $\tau^i(x) = \eta^i x$, hence x has d distinct conjugates and $L = K[x]$. Furthermore

$$x^d = \prod_{i=1}^{d} \eta^i x = \prod_{i=0}^{d} \tau^i(x) = N_{L/K}(x) \in K.$$

$\square$

Proposition 9.50 *There exists $x \in L$, $x \neq 0$, such that $\tau(x) = \eta x$.*

Proof Note first that $\eta^{-1} \in K$ implies $\tau(\eta^{-1}) = \eta^{-1}$, hence $\prod_{i=1}^{d} \tau^i(\eta^{-1}) = (\eta^{-1})^d = 1$. Consider now an element $a \in L$ such that $\prod_{i=1}^{d} \tau^i(a) = 1$. We want to prove that there exists $b \in L$ such that $a\tau(b) = b$.

Put $e_i = a\tau(a)...\tau^i(a)$. We claim that

$$a\tau(\sum_{i=0}^{d-1} e_i \tau^i(c)) = \sum_{i=0}^{d-1} e_i \tau^i(c)$$

for all $c \in L$. Indeed

$$a\tau(e_i \tau^i(c)) = e_{i+1}\tau^{i+1}(c) \quad \text{for} \quad 0 \le i < d-1 \quad \text{and}$$

$$a\tau(e_{d-1}\tau^{d-1}(c)) = a \prod_{i=1}^{d} \tau^i(a)\tau^d(c) = ac.$$

It is now sufficient to show that there exists $c \in L$ such that $\sum_{i=0}^{d-1} e_i \tau^i(c) \neq 0$. But this is a consequence of the following general lemma.

Lemma 9.51 *(Dedekind) Let $K \subset L$ be an extension. Let $g_1, ..., g_n$ be distinct K-isomorphisms from L to Ω. If $z_i \in \Omega$, with $i = 1, ..., n$, are such that $\sum_{i=1}^{n} z_i g_i = 0$, then $z_i = 0$ for $i = 1, ..., n$.*

Proof We use induction on n. The result is obvious for $n = 1$. Assume $\sum_{i=1}^{n} z_i g_i = 0$. If $x \in L$ and $y \in L$, we have

$$\sum_{i=1}^{n} z_i g_i(xy) = \sum_{i=1}^{n} z_i g_i(x) g_i(y) = 0,$$

which implies, for all $x \in L$, that

$$\sum_{i=1}^{n} z_i g_i(x) g_i = 0.$$

Substracting $g_n(x) \sum_{i=1}^{n} z_i g_i = 0$, we find a new relation

$$\sum_{i=1}^{n-1} z_i (g_i(x) - g_n(x)) g_i = 0,$$

yielding $z_i(g_i(x) - g_n(x)) = 0$ for $i = 1, .., n - 1$. Since $g_i \neq g_n$, for $i < n$, there exists $x \in L$ such that $g_i(x) \neq g_n(x)$. As a consequence $z_i = 0$, for $i < n$, and $z_n = 0$. We are done. $\qquad\square$

9.6   Exercises

Here, all fields have characterisic zero.

1. Let $K \subset L$ be a finite extension of fields. Assume that $[L : K]$ is a prime number. If $x \in L$ and $x \notin K$, show that $L = K[x]$.

2. Let Ω be an algebraic closure of K. Consider L and L' two finite extensions of K contained in Ω and F the smallest extension of K containing L and L'. Find a relation between $[F : K]$, $[L : K]$, $[L' : K]$ and $[L \cap L' : K]$.

3. Consider $f \in K[X]$ and L the decomposition field of f in an algebraic closure of K. Show that the Galois group of the normal extension $K \subset L$ acts transitively on the roots of f if and only if f is irreducible.

4. Consider $K \subset L$ a finite extension of fields, Ω an algebraic closure of L and $F \subset \Omega$ the smallest normal extension of K containing L. Assume that $x \in F$ is such that the conjugates of x form a basis of F over K. Show that $L = K[\text{Tr}_{F/L}(x)]$.

5. Consider the involution $i(P(X)) = P(-X)$ on the field $K(X)$. If G is the group, with two elements, generated by this involution, show $K(X)^G = K(X^2)$.

6. Consider the natural action of the permutation group S_n on the field $K(X_1, \ldots, X_n)$. Put $L = K(X_1, \ldots, X_n)^{S_n}$. If

$$\prod_{i=1}^{n}(Z - X_i) = Z^n + s_1 Z^{n-1} + \cdots + s_n,$$

show first that $s_i \in L$ for $i = 1, \ldots, n$ and next that $L = K[s_1, \ldots, s_n]$. Hint: prove $[L : K[s_1, \ldots, s_n]] \leq n!$.

7. Let Ω be an algebraic closure of K and $x \in \Omega$. If $L \subset \Omega$ is the smallest normal extension of K such that $x \in L$, give an upper bound, depending on $[K(x) : K]$, for $[L : K]$.

8. Consider the field $L = \mathbb{Q}[i, e^{2i\pi/5}] \subset \mathbb{C}$. Show that $L = \mathbb{Q}[e^{2i\pi/20}]$, that L is the field of decomposition of $X^{20} - 1 = 0$ and compute $[L : \mathbb{Q}]$.

   Consider the $\mathbb{Q}$-automorphisms σ and τ of L defined by

   $$\sigma(e^{2i\pi/20}) = e^{6i\pi/20} \quad \text{and} \quad \tau(e^{2i\pi/20}) = e^{38i\pi/20}.$$

   Show that σ and τ commute and that the group homomorphism $\psi : \mathbb{Z} \times \mathbb{Z} \to \mathrm{Gal}(L/\mathbb{Q})$ defined by $\psi(m, n) = \sigma^m \tau^n$ induces an isomorphism $(\mathbb{Z}/4\mathbb{Z}) \times (\mathbb{Z}/2\mathbb{Z}) \simeq \mathrm{Gal}(L/\mathbb{Q})$.

10

Noether's normalization lemma

In this chapter, we study finitely generated algebras over an infinite field. Several of our statements are also true for finitely generated algebras over any field. When the generalization to finite fields is essentially straightforward, we do it. But we want to move fast towards complex geometry and to this end we choose the most efficient version of the normalization lemma. This form requires infinitely many elements in the field. We will not bother to adapt our proofs to the most general situation. The reader interested in using finite fields will have to do a bit of work on his own, or consult another text.

10.1   Transcendence degree

Definition 10.1 *Let $A \subset R$ be a ring extension. We say that $x_1, ..., x_s \in R$ are algebraically independent (or algebraically free) over A if the ring homomorphism*

$$A[X_1, ..., X_s] \to A[x_1, ..., x_s] \qquad P \to P(x_1, ..., x_n)$$

is an isomorphism.

Definition 10.2 *Let $K \subset L$ be a field extension and $x_1, ..., x_n \in L$ be algebraically independent over K. If L is algebraic on the fraction field $K(x_1, ..., x_n)$ of $K[x_1, ..., x_n]$, we say that $(x_1, ..., x_n)$ is a transcendence basis of L over K.*

Definition 10.3 *Let $K \subset L$ be a field extension. If there exist $z_1, ..., z_n \in L$ such that L is the fraction field $K(z_1, ..., z_n)$ of $K[z_1, ..., z_n]$, we say that L is a finitely generated field extension of K.*

The proof of the next result is straightforward and left to the reader.

Proposition 10.4 *Let $K \subset L$ be a field extension. If $x_1, ..., x_n \in L$ are such that L is algebraic over the fraction field $K(x_1, ..., x_n)$ of $K[x_1, ..., x_n]$, there exists a transcendence basis extracted from $(x_1, ..., x_n)$.*

In particular a finitely generated field extension of K has a transcendence basis over K.

Theorem 10.5 *Let $K \subset L$ be a field extension such that L has a transcendence basis with n elements over K. Then*

(i) *if $x_1, ..., x_n \in L$ are algebraically independent over K, then $(x_1, ..., x_n)$ is a transcendence basis of L over K;*

(ii) *all transcendence bases of L over K have n elements.*

Proof We can assume that all transcendence bases of L over K have at least n elements. We proceed by induction over n. The assertion is obvious for $n = 0$.

If $n > 0$, let $(z_1, ..., z_n) \in L$ be a transcendence basis of L over K. Since x_1 is algebraic over $K(z_1, ..., z_n)$, there exists a non-zero polynomial $f \in K[Z_1, ..., Z_n, X_1]$ such that $f(z_1, ..., z_n, x_1) = 0$. Since x_1 is transcendent over K, the polynomial f has to depend on some z_i, for example z_1. This shows that z_1 is algebraic over $K(z_2, ..., z_n, x_1)$, hence that L is algebraic over $K(z_2, ..., z_n, x_1)$ by Corollary 8.15. Since n is minimal this proves that $(z_2, ..., z_n, x_1)$ is a transcendence basis of L over K. As a consequence, $z_2, ..., z_n$ is a transcendence basis of L over $K(x_1)$, and we conclude by induction. □

Definition 10.6 *Let $K \subset L$ be a field extension.*

(i) *If L has a transcendence basis with n elements, over K, we say that the field L has transcendence degree n over K and we write $\mathrm{trdeg}_K(L) = n$.*

(ii) *If L has no finite transcendence basis over K, we write $\mathrm{trdeg}_K(L) = \infty$.*

(iii) *If $A \subset B$ is an extension of domains and $K(A) \subset K(B)$ the corresponding extension of fraction fields, we define $\mathrm{trdeg}_A(B) = \mathrm{trdeg}_{K(A)}(K(B))$.*

10.2   The normalization lemma

In this short section we only prove the normalization lemma. This fundamental result has so many algebraic and geometric consequences that we cannot hope to explain them in a few words. We can only advise our reader to keep this result in mind constantly, particularly when the time comes to think about schemes and varieties.

Theorem 10.7 *(The normalization lemma) Let $A = K[x_1, \dots, x_n]$ be a finitely generated algebra on an infinite field K. There exist $y_1, \dots, y_r \in A$ such that:*

(i) *the elements $y_1, \dots, y_r$ are algebraically independent over K;*

(ii) *the ring A is finite over $K[y_1, \dots, y_r]$;*

(iii) *the elements $y_1, \ldots, y_r$ are linear combinations of $x_1, \ldots, x_n$, with coeffi-cients in K.*

Our proof depends on the following result.

Lemma 10.8 *Let K be an infinite field and $F \in K[X_1, \ldots, X_n]$ a non-zero polynomial. There exist $(a_1, \ldots, a_n) \in K^n$ such that $F(a_1, \ldots, a_n) \neq 0$.*

Proof This assertion is well known for $n = 1$. We prove it by induction on n. We can assume that F depends on the variable X_n and order F with regard to it :

$$F = G_0(X_1, \ldots, X_{n-1})X_n^r + \cdots + G_r(X_1, \ldots, X_{n-1}),$$

with $r > 0$ and $G_0(X_1, \ldots, X_{n-1}) \neq 0$. By the induction hypothethis, there exists $(a_1, \ldots, a_{n-1}) \in K^{n-1}$ such that $G_0(a_1, \ldots, a_{n-1}) \neq 0$. The polynomial $G_0(a_1, \ldots, a_{n-1})X_n^r + \cdots + G_r(a_1, \ldots, a_{n-1})$, in the variable X_n, is non-zero. Hence there exists $a_n \in K$ which is not a root of it. □

Definition 10.9 *A polynomial $F \in K[X_1, \ldots, X_n]$ is homogeneous of degree r if it is a combination of monomials of degree r.*

Consider a monomial $M(X_1, \ldots, X_n) = X_1^{l_1} \cdots X_n^{l_n}$, with $\sum l_i = r$. If $a \in K$, it is clear that $M(aX_1, \ldots, aX_n) = a^r M(X_1, \ldots, X_n)$. Consequently, if F is homogeneous of degree r , then

$$F(aX_1, \ldots, aX_n) = a^r F(X_1, \ldots, X_n).$$

We begin by proving a special case of the normalization lemma.

Lemma 10.10 *Let K be an infinite field and $f \in K[X_1, \ldots, X_n]$ a non-zero, non-constant polynomial. There exist $Z_1, \ldots, Z_{n-1} \in K[X_1, \ldots, X_n]$ such that:*

(i) *the elements $Z_1, \ldots, Z_r$ are linear combinations of $X_1, \ldots, X_n$;*

(ii) *the classes $\overline{Z}_i = \mathrm{cl}(Z_i) \in K[X_1, \ldots, X_n]/(f)$ are algebraically indepen-dent;*

(iii) *the ring $K[X_1, \ldots, X_n]/(f)$ is finite over $K[\overline{Z}_1, \ldots, \overline{Z}_{n-1}]$.*

Proof Consider a decomposition $f = F_r + \cdots + F_0$, with F_i homogeneous of degree i and $F_r \neq 0$. There exists $(b_1, \ldots, b_n) \in K^n$ such that $F_r(b_1, \ldots, b_n) \neq 0$. Since F_r is homogeneous, we have $(b_1, \ldots, b_n) \neq (0, \ldots, 0) \in K^n$. We can assume $b_n \neq 0$, for example. Putting $b_i/b_n = a_i$, we have $F_r(a_1, \ldots, a_{n-1}, 1) \neq 0$.

If $Z_i = X_i - a_i X_n$, for $1 \leq i \leq (n-1)$, we have clearly $K[X_1, \ldots, X_n] = K[Z_1, \ldots, Z_{n-1}, X_n]$. Note that

$$
\begin{aligned}
F_i(X_1, \ldots, X_n) &= F_i(Z_1 + a_1 X_n, \ldots, Z_{n-1} + a_{n-1} X_n, X_n) \\
&= F_i(a_1, \ldots, a_{n-1}, 1) X_n^i + G_i(Z_1, \ldots, Z_{n-1}, X_n),
\end{aligned}
$$

where G_i is a polynomial of degree i whose degree in the variable X_n is strictly less than i. This shows

$$
f(X_1, \ldots, X_{n-1}, X_n) = F_r(a_1, \ldots, a_{n-1}, 1) X_n^r + g(Z_1, \ldots, Z_{n-1}, X_n),
$$

where $\deg_{X_n}(g) < r$. We have proved that $K[X_1, \ldots, X_n]/(f)$ is finite over $K[\overline{Z}_1, \ldots, \overline{Z}_{n-1}]$.

To conclude, we must show that the elements

$$
\overline{Z}_1, \ldots, \overline{Z}_{n-1} \in K[X_1, \ldots, X_n]/(f)
$$

are algebraically independent. Consider $h \in K[Z_1, \ldots, Z_{n-1}]$ such that

$$
h(\overline{Z}_1, \ldots, \overline{Z}_{n-1}) = 0.
$$

This implies

$$
h \in fK[Z_1, \ldots, Z_{n-1}, X_n] \subset K[Z_1, \ldots, Z_{n-1}, X_n].
$$

Since h does not depend on the variable X_n on which f depends, we have $h = 0$. $\qquad\square$

Proof of the normalization lemma by induction on n:

If $x_1, \ldots, x_n$ are algebraically independent over K, there is nothing to prove. If not, let us find linear combinations $x_1', \ldots, x_{n-1}'$ of $x_1, \ldots, x_n$, such that A is finite over $K[x_1', \ldots, x_{n-1}']$.

Consider a relation $f(x_1, \ldots, x_n) = 0$, where $f(X_1, \ldots, X_n)$ is a non-zero polynomial. By Lemma 10.10, there exist linear combinations $X_1', \ldots, X_{n-1}'$ of $X_1, \ldots, X_n$, such that $K[X_1, \ldots, X_n]/(f)$ is finite over $K[\overline{X'}_1, \ldots, \overline{X'}_{n-1}]$, where

$$
\overline{X'}_i = \mathrm{cl}(X_i') \in K[X_1, \ldots, X_n]/(f).
$$

Putting $x_i' = \mathrm{cl}(X_i') \in K[x_1, \ldots, x_n]$, it is clear that $K[x_1, \ldots, x_n]$ is finite over $K[x_1', \ldots, x_{n-1}']$.

Now, by the induction hypothesis, there exist linear combinations $y_1, \ldots, y_r$ of $x_1', \ldots, x_{n-1}'$, such that $y_1, \ldots, y_r$ are algebraically independent over K and that $K[x_1', \ldots, x_{n-1}']$ is finite over $K[y_1, \ldots, y_r]$. We are done.

10.3   Hilbert's Nullstellensatz

We have already proved Hilbert's Nullstellensatz as an exercise; see Exercise 8.25. We give a new proof here. More precisely, we show that this result is a consequence of the normalization lemma. We believe furthermore that this new proof enlightens the meaning of the Nullstellensatz.

Theorem 10.11 *(Hilbert's Nullstellensatz)*
Let $K \subset L$ be a field extension. If L is a finitely generated K-algebra then L is finite over K.

Proof   If K is infinite, there exist an integer $r \geq 0$ and elements $y_1, \ldots, y_r \in L$, algebraically independent over K, such that L is finite over $K[y_1, \ldots, y_r]$. Since L is a field, the polynomial ring $K[y_1, \ldots, y_r]$ is a field by Theorem 8.21. This implies $r = 0$, and L is finite over K.

If K is finite, consider an isomorphism $L \simeq K[X_1, \ldots, X_n]/\mathcal{M}$, where $\mathcal{M}$ is a maximal ideal of the polynomial ring $K[X_1, \ldots, X_n]$. Let K' be an algebraic closure of the field K. The polynomial ring $K'[X_1, \ldots, X_n]$ is obviously integral over $K[X_1, \ldots, X_n]$. Hence there exists a maximal ideal $\mathcal{N} \subset K'[X_1, \ldots, X_n]$ such that $\mathcal{N} \cap K[X_1, \ldots, X_n] = \mathcal{M}$. Since K' is infinite, the K'-algebra $K'[X_1, \ldots, X_n]/\mathcal{N}$ is finite over K', hence algebraic over K. Using the double inclusion $K \subset K[X_1, \ldots, X_n]/\mathcal{M} \subset K'[X_1, \ldots, X_n]/\mathcal{N}$, we see that the field $K[X_1, \ldots, X_n]/\mathcal{M}$ is algebraic over K. Now if we put $x_i = \mathrm{cl}(X_i) \in K[X_1, \ldots, X_n]/\mathcal{M}$, then $L = K[x_1, \ldots, x_n]$, where x_i is finite over K. Obviously, this implies that L is finite over K.   □

Before presenting and discussing the main applications of the Nullstellensatz, we recall a few facts:

- to each point $a = (a_1, \ldots, a_n) \in K^n$ we can associate the evaluation homomorphism

$$e_a : K[X_1, \ldots, X_n] \to K, \qquad e_a(P) = P(a_1, \ldots, a_n)$$

  whose kernel is, as we know, the maximal ideal $(X_1 - a_1, \ldots, X_n - a_n)$ of $K[X_1, \ldots, X_n]$;

- if $\mathcal{I}$ is an ideal of $K[X_1, \ldots, X_n]$, the closed set

$$V(\mathcal{I}) \subset \mathrm{Spec}(K[X_1, \ldots, X_n])$$

  is

$$V(\mathcal{I}) = \{\mathcal{P} \in \mathrm{Spec}(K[X_1, \ldots, X_n]), \quad \mathcal{I} \subset \mathcal{P}\};$$

- we denote by $\mathrm{Spec_m}(A) \subset \mathrm{Spec}(A)$ the set of maximal ideals of a ring A. The sets

$$
\begin{aligned}
V_{\mathrm{m}}(\mathcal{I}) &= V(\mathcal{I}) \cap \mathrm{Spec_m}(K[X_1, \ldots, X_n]) \\
&= \{\mathcal{M} \in \mathrm{Spec_m}(K[X_1, \ldots, X_n]), \quad \mathcal{I} \subset \mathcal{M}\}
\end{aligned}
$$

are the closed sets of $\mathrm{Spec_m}(K[X_1, \ldots, X_n])$ for the topology induced by the Zariski topology and also called the Zariski topology.

Corollary 10.12 *Let K be an algebraically closed field.*

(i) *The correspondence*

$$
a = (a_1, \ldots, a_n) \to \ker(e_a) = (X_1 - a_1, \ldots, X_n - a_n)
$$

between K^n and $\mathrm{Spec_m}(K[X_1, \ldots, X_n])$ is bijective.

(ii) *If $\mathcal{I}$ is an ideal of $K[X_1, \ldots, X_n]$, this correspondence induces a bijective correspondence between the set of points $a = (a_1, \ldots, a_n) \in K^n$ such that $F(a_1, \ldots, a_n) = 0$ for all $F \in \mathcal{I}$ and the closed set $V_{\mathrm{m}}(\mathcal{I})$ of $\mathrm{Spec_m}(K[X_1, \ldots, X_n])$.*

Proof Note first that $(a_1, \ldots, a_n) \neq (b_1, \ldots, b_n)$ implies obviously $\ker(e_a) \neq \ker(e_b)$.

Let $\mathcal{M}$ be a maximal ideal of $K[X_1, \ldots, X_n]$. We want to show that there exists a point $(a_1, \ldots, a_n) \in K^n$ such that $\mathcal{M} = \ker(e_a)$. The field $K[X_1, \ldots, X_n]/\mathcal{M}$ is a finitely generated K-algebra. Hence it is algebraic over K. Since K is algebraically closed, the composed homomorphism

$$
K \to K[X_1, \ldots, X_n] \to K[X_1, \ldots, X_n]/\mathcal{M}
$$

is an isomorphism (Theorem 10.11). As a consequence, there exists, for each i, an element $a_i \in K$ such that $\mathrm{cl}(X_i) = \mathrm{cl}(a_i) \in K[X_1, \ldots, X_n]/\mathcal{M},$. This proves $(X_1 - a_1, \ldots, X_n - a_n) \subset \mathcal{M}$, hence $\mathcal{M} = \ker(e_a)$.

Now if $\mathcal{M} = \ker(e_a)$, it is clear that $\mathcal{I} \subset \mathcal{M}$ if and only if $e_a(F) = 0$ for all $F \in \mathcal{I}$ and we are done. $\qquad\square$

Let K be an algebraically closed field. From Corollary 10.12, we see that the Zariski topology on $\mathrm{Spec_m}(K[X_1, \ldots, X_n])$ induces a topology on K^n, once more called the Zariski topology, whose closed sets are called the closed algebraic sets of K^n. In a bold and courageous move we decide to denote the closed algebraic set of K^n corresponding to $V_{\mathrm{m}}(\mathcal{I})$ by $V(\mathcal{I}) \subset K^n$. In other words,

$$
V(\mathcal{I}) = \{(a_1, \ldots, a_n) \in K^n, \quad F(a_1, \ldots, a_n) = 0 \text{ for all } F \in \mathcal{I}\}.
$$

This seems an abuse of notation. I would like to convince you that it is not. We have proved that the closed sets $V_m(\mathcal{I}) \subset \operatorname{Spec}_m(K[X_1, ..., X_n])$ and $V(\mathcal{I}) \subset K^n$ are in natural bijective correspondence. To conclude, we prove that $V_m(\mathcal{I}) \subset \operatorname{Spec}_m(K[X_1, ..., X_n])$ and $V(\mathcal{I}) \subset \operatorname{Spec}(K[X_1, ..., X_n])$ do characterize each other. In other words, we must be able to recover $V(\mathcal{I}) \subset \operatorname{Spec}(K[X_1, ..., X_n])$ from $V_m(\mathcal{I}) = V(\mathcal{I}) \cap \operatorname{Spec}_m(K[X_1, ..., X_n])$. Since, in $\operatorname{Spec}(K[X_1, ..., X_n])$, we have $V(\mathcal{I}) = V(\sqrt{\mathcal{I}})$, this can be done by using the next theorem, often also called the Nullstellensatz.

Theorem 10.13 *Let K be a field and $\mathcal{I} \subset K[X_1, ..., X_n]$ an ideal. Then*

$$\sqrt{\mathcal{I}} = \bigcap_{\mathcal{M} \in V_m(\mathcal{I})} \mathcal{M}.$$

Proof The inclusion

$$\sqrt{\mathcal{I}} \subset \cap_{\mathcal{M} \in V_m(\mathcal{I})} \mathcal{M}$$

is obvious. If $g \notin \sqrt{\mathcal{I}}$, we show that there exists $\mathcal{N} \in V_m(\mathcal{I})$ such that $g \notin \mathcal{N}$. To this end, consider the multiplicatively stable part $S = \{g^n\}_{n \geq 0}$ and note that $S \cap \mathcal{I} = \emptyset$. This implies

$$\mathcal{I}K[X_1, ..., X_n]_g \neq K[X_1, ..., X_n]_g.$$

Let $\mathcal{M}$ be a maximal ideal of the ring $K[X_1, ..., X_n]_g$ containing the ideal $\mathcal{I}K[X_1, ..., X_n]_g$. Note that $K[X_1, ..., X_n]_g \simeq K[X_1, ..., X_n, g^{-1}]$ is a finitely generated K-algebra, hence that the field $K[X_1, ..., X_n]_g/\mathcal{M}$ is one also. By Theorem 10.11, the field extension $K \subset K[X_1, ..., X_n]_g/\mathcal{M}$ is algebraic.

Put $\mathcal{N} = \mathcal{M} \cap K[X_1, ..., X_n]$. Since $\mathcal{I}K[X_1, ..., X_n]_g \subset \mathcal{M}$, it is clear that $\mathcal{I} \subset \mathcal{N}$. Furthermore $S \cap \mathcal{N} = \emptyset$. The double inclusion

$$K \subset K[X_1, ..., X_n]/\mathcal{N} \subset K[X_1, ..., X_n]_g/\mathcal{M}$$

shows that the ring $K[X_1, ..., X_n]/\mathcal{N}$ is an integral extension of K. Consequently, this ring is a field and $\mathcal{N}$ is a maximal ideal of $K[X_1, ..., X_n]$. Hence we found a maximal ideal $\mathcal{N}$ such that $\mathcal{I} \subset \mathcal{N}$ and $g \notin \mathcal{N}$. Our theorem is proved. $\square$

Exercise 10.14 Let $f \in \mathbb{C}[X, Y]$ be such that $f(a, a) = f(a, -a) = 0$ for all $a \in \mathbb{C}$. Show that $f \in (X^2 - Y^2)\mathbb{C}[X, Y]$.

10.4   Jacobson rings

An ideal is said to be *radical* if it is equal to its radical, in other words if it is an intersection of prime ideals.

Definition 10.15 *A ring in which every radical ideal is an intersection of maximal ideals is a Jacobson ring.*

We note that A is a Jacobson ring if and only if every prime ideal is an intersection of maximal ideals. As an obvious consequence of Theorem 10.13, we see that a finitely generated algebra over a field is a Jacobson ring.

We stress once more the following essential property of Jacobson rings.

Proposition 10.16 *Let A be a Jacobson ring and $F \subset \mathrm{Spec}(A)$ a closed set. Then F is characterized by $F \cap \mathrm{Spec}_{\mathrm{m}}(A)$. More precisely,*

$$F = V(\bigcap_{\mathcal{M} \in F \cap \mathrm{Spec}_{\mathrm{m}}(A)} \mathcal{M}).$$

This is an obvious consequence of the definition.

Exercise 10.17 Let A be a Jacobson ring and let $\mathcal{M}$ be a maximal ideal of $A[X_1, \ldots, X_n]$. Show that $\mathcal{M} \cap A$ is a maximal ideal of A.

10.5   Chains of prime ideals in geometric rings

A chain of length l, of prime ideals of a ring A, is a strictly increasing sequence $\mathcal{P}_0 \subset \ldots \subset \mathcal{P}_l$ of prime ideals of A. Note that there are in fact $l + 1$ prime ideals in a chain of length l.

Theorem 10.18 *Let K be a field, let A be a finitely generated K-algebra and let $x_1, \ldots, x_n \in A$ algebraically independent over K such that A is integral over $K[x_1, \ldots, x_n]$.*

(i) *If $\mathcal{P}_0 \subset \ldots \subset \mathcal{P}_l$ is a chain of prime ideals of A, then $l \leq n$.*

(ii) *If the chain $\mathcal{P}_0 \subset \ldots \subset \mathcal{P}_l$ cannot be extended, then*

$$l = n \iff \mathcal{P}_0 \cap K[x_1, \ldots x_n] = (0).$$

Proof If $n = 0$, the ring A is Artinian and (i) and (ii) are obvious. We proceed by induction on n.

If $\mathcal{Q}_i = \mathcal{P}_i \cap K[x_1, ..., x_n]$, then $\mathcal{Q}_0 \subset ... \subset \mathcal{Q}_l$ is a chain of prime ideals of $K[x_1, ..., x_n]$ by Theorem 8.26 (ii). Since $\mathcal{Q}_1 \neq (0)$, there exists a non-zero element $f \in \mathcal{Q}_1$. Then $\mathcal{Q}_1/(f) \subset ... \subset \mathcal{Q}_l/(f)$ is a chain of prime ideals, of length $(l - 1)$, of the ring $K[x_1, ..., x_n]/(f)$. By Lemma 10.10, this ring is integral over a polynomial ring $K[Y_1, ..., Y_{n-1}]$. Hence, we have $(l-1) \leq (n-1)$, by the induction hypothesis.

Assume now that the chain $(\mathcal{P}_i)_{0 \leq i \leq l}$ is not extendable. If $\mathcal{Q}_0 = (0)$, put $A' = A/\mathcal{P}_0$ and $\mathcal{P}'_i = \mathcal{P}_i/\mathcal{P}_0$. The domain A' is integral over its subring $K[x_1, ..., x_n]$ and we can apply the going-down theorem to this extension of rings.

Since $K[x_1, ..., x_n]$ is a UFD, there exists an irreducible element $f \in \mathcal{Q}_1$. Note that $fK[x_1, ..., x_n]$ is a prime ideal. Since $fK[x_1, ..., x_n] \subset \mathcal{Q}_1 = \mathcal{P}'_1 \cap K[x_1, ..., x_n]$, there exists a prime ideal $\mathcal{N} \subset \mathcal{P}'_1$ of A' such that

$$\mathcal{N} \cap K[x_1, ..., x_n] = fK[x_1, ..., x_n].$$

But the chain $\mathcal{P}'_0 \subset \mathcal{P}'_1$ is not extendable and $\mathcal{P}'_0 = (0)$, hence $\mathcal{N} = \mathcal{P}'_1$, and $fK[x_1, ..., x_n] = \mathcal{Q}_1$. This shows that $K[x_1, ..., x_n]/(f)$ is a subring of $A/\mathcal{P}_1$ on which this domain is integral. Now we recall that, by Lemma 10.10, $K[x_1, ..., x_n]/(f)$ is integral over a polynomial subring $K[Y_1, ..., Y_{n-1}]$. Since $(0) = \mathcal{P}_1/\mathcal{P}_1 \subset ... \subset \mathcal{P}_l/\mathcal{P}_1$ is a non-extendable chain of prime ideals of $A/\mathcal{P}_1$ such that $(0) \cap K[Y_1, ..., Y_{n-1}] = (0)$, we have $l - 1 = n - 1$, by the induction hypothesis.

If $\mathcal{Q}_0 \neq (0)$, let $g \in \mathcal{Q}_0$ be non-zero. By Lemma 10.10, the ring $K[x_1, ..., x_n]/(g)$ is integral over a polynomial ring in $n - 1$ variables. Hence the chain

$$\mathcal{Q}_0/(g) \subset ... \subset \mathcal{Q}_l/(g)$$

has length at most $n - 1$ and $l < n$. $\qquad\square$

Corollary 10.19 *Let K be an infinite field and A a domain finitely generated as a K-algebra. Then all non-extendable chains of prime ideals of A have length $\mathrm{trdeg}_K(K(A))$.*

Proof Put $n = \mathrm{trdeg}_K(K(A))$. By the normalization lemma, there exist algebraically independent elements $x_1, ..., x_n \in A$ such that A is integral over $K[x_1, ..., x_n]$. Now, since A is a domain, for any non-extendable chain of prime ideals $\mathcal{P}_0 \subset ... \subset \mathcal{P}_l$, we have $\mathcal{P}_0 = (0)$, hence $\mathcal{P}_0 \cap K[x_1, ..., x_n] = (0)$. We conclude by Theorem 10.18 $\qquad\square$

As an immediate consequence of this corollary we get the following.

Corollary 10.20 *Let K be an infinite field, A a domain finitely generated as a K-algebra and $Q \subset P$ two prime ideals of A. If*

$$Q = P_0 \subset P_1 \subset \ldots \subset P_l = P$$

is a non-extendable chain of prime ideals between Q and P, then

$$l = \operatorname{trdeg}_K(K(A/Q)) - \operatorname{trdeg}_K(K(A/P)).$$

10.6   Height and dimension

In this section, we come back to general Noetherian rings. We study their chains of prime ideals.

Proposition 10.21 *Let A be a Noetherian ring, $\mathcal{I} = (a_1, \ldots, a_n)$ an ideal generated by n elements and P a minimal prime ideal of $\mathcal{I}$. If $P_0 \subset \ldots \subset P_l = P$ is a chain of prime ideals, then $l \leq n$.*

To begin with, we prove this result for $n = 1$, when A is a domain.

Lemma 10.22 *Let A be a Noetherian domain and $a \in A$ a non-zero element. If P is a minimal prime ideal of aA, then (0) is the only prime ideal strictly contained in P.*

Proof   By considering the local ring A_P we can assume that A is a local domain with maximal ideal P and show that P is the only non-zero prime ideal.

Let Q be a prime ideal strictly contained in P. We want to prove $Q = (0)$. Consider the decreasing sequence of ideals $Q^{(n)} + aA$. Note that since A/aA has only one prime ideal, it is necessarily Artinian. Now, since $aA \subset Q^{(n)} + aA$ for all n, there exists m such that

$$Q^{(n)} + aA = Q^{(m)} + aA \quad \text{for } n \geq m.$$

This shows $Q^{(m)} \subset Q^{(n)} + aA$ for $n > m$. Since $Q^{(m)}$ is Q-primary and $a \notin Q$,

$$ab \in Q^{(m)} \Rightarrow b \in Q^{(m)},$$

hence

$$Q^{(m)} \subset Q^{(n)} + aQ^{(m)} \quad \text{for } n \geq m.$$

But a is contained in the Jacobson radical of A, hence $Q^{(m)} = Q^{(n)}$ for all $n \geq m$, by Nakayama's lemma. Note then that $Q^{(n)} \subset Q^n A_Q$, since A is a domain. This shows

$$Q^{(m)} = \cap_{n \geq m} Q^{(n)} \subset \cap_{n \geq m} Q^n A_Q.$$

But $\cap_{n \geq m} \mathcal{Q}^n A_\mathcal{Q} = (0)$, by Krull's theorem, hence we have proved $\mathcal{Q}^{(m)} = (0)$. Since A is a domain, this implies $\mathcal{Q} = (0)$. $\qquad\qquad\qquad\square$

Proof of Proposition 10.21 We proceed now by induction on n. Replacing if necessary the ring A, by the ring $A_\mathcal{P}$, we can assume that A is local with maximal ideal $\mathcal{P}$. Furthermore, we can also assume that $\mathcal{P}_{l-1}$ is maximal in the set of prime ideals strictly contained in $\mathcal{P}$ and that $a_n \notin \mathcal{P}_{l-1}$. From this, we deduce that $\mathcal{P}_{l-1} + a_n A$ is $\mathcal{P}$-primary, and consequently that there exists k such that $a_i^k \in \mathcal{P}_{l-1} + a_n A$ for all $i \in [1, n-1]$. Put $a_i^k = b_i + a_n c_i$, with $b_i \in \mathcal{P}_{l-1}$. Obviously, we have $(a_1^k, ..., a_{n-1}^k, a_n) = (b_1, ..., b_{n-1}, a_n)$. This shows that $\mathcal{P}$ is a minimal prime ideal of $(b_1, ..., b_{n-1}, a_n)$. We claim that $\mathcal{P}_{l-1}$ is a minimal prime ideal of $(b_1, ..., b_{n-1})$. Indeed, let $\mathcal{Q}$ be a minimal prime ideal of $(b_1, ..., b_{n-1})$ contained in $\mathcal{P}_{l-1}$. Since $\mathcal{P}/\mathcal{Q}$ is a minimal prime ideal of $(\mathcal{Q} + a_n A)/\mathcal{Q}$, we have $\mathcal{P}_{l-1}/\mathcal{Q} = (0)$, by Lemma 10.22. By the induction hypothesis, we find $l - 1 \leq n - 1$ and we are done.

Corollary 10.23 *Let A be a Noetherian ring and $\mathcal{P} = (a_1, ..., a_r)$ a prime ideal. A chain of prime ideals all contained in $\mathcal{P}$ has length at most r.*

Proof This is an immediate consequence of Proposition 10.21. $\qquad\qquad\square$

Example 10.24 In the polynomial ring $K[X_1, ..., X_n]$ over a field K, the ideal $(X_1, ..., X_n)$ is generated by n elements and $(0) \subset (X_1) \subset (X_1, X_2) \subset \cdots \subset (X_1, ..., X_n)$ is a chain of length n, of prime ideals all contained in $(X_1, ..., X_n)$.

We can now define the dimension of a Noetherian ring and the height of a prime ideal.

Definition 10.25

(i) *The dimension $\dim(R)$ of a local Noetherian ring R is the largest integer n such that there exists a chain of length n, of prime ideals of this ring.*

(ii) *The height $\operatorname{ht}(\mathcal{P})$ of a prime ideal $\mathcal{P}$, of a Noetherian ring A, is the dimension of the local ring $A_\mathcal{P}$.*

(iii) *The dimension of a Noetherian ring A is $\dim(A) = \sup(\operatorname{ht}(\mathcal{P}))$, for $\mathcal{P} \in \operatorname{Spec}(A)$.*

Note that $\operatorname{ht}(\mathcal{P}) = \dim(A_\mathcal{P})$ is finite, by Proposition 10.21 or Corollary 10.23.

We state and prove two other results concerning all Noetherian rings.

Theorem 10.26 *Let $\mathcal{P}$ be a prime ideal of a Noetherian ring A. Then $\mathrm{ht}(\mathcal{P})$ is the smallest integer h for which there exist elements $a_1, ..., a_h \in \mathcal{P}$ such that $\mathcal{P}$ is a minimal prime ideal of $(a_1, ..., a_h)$.*

Proof We know that if $\mathcal{P}$ is a minimal prime ideal of $(a_1, ..., a_n)$, then $\mathrm{ht}(\mathcal{P}) \leq n$. We put $h = \mathrm{ht}(\mathcal{P})$ and we prove by induction on h that there exist $a_1, ..., a_h \in \mathcal{P}$ such that $\mathcal{P}$ is a minimal prime ideal of $(a_1, ..., a_h)$.

This is obvious for $h = 0$. Assume $h > 0$. Let $\mathcal{Q}_i$, with $i = 1, ..., k$, be the minimal prime ideal of A contained in $\mathcal{P}$. By the avoiding lemma, there exists $f \in \mathcal{P}$ such that $f \notin \mathcal{Q}_i$ for $i = 1, ..., k$. Clearly, the height h' of the prime ideal $\mathcal{P}/fA$ of the ring A/fA is less than or equal to $h - 1$. By the induction hypothesis, there exist elements $b_1, ..., b_{h'} \in A/fA$ such that $\mathcal{P}/fA$ is a minimal prime ideal of $(b_1, ..., b_{h'})$. If $a_i \in A$ is such that $\mathrm{cl}(a_i) = b_i \in A/fA$, then $\mathcal{P}$ is a minimal prime ideal of $(a_1, ..., a_{h'}, f)$. Hence $h = \mathrm{ht}(\mathcal{P}) \leq h' + 1$. This implies $h = h' + 1$ and we are done. $\square$

Corollary 10.27 *Let A and B be Noetherian local rings and $\mathcal{M}$ and $\mathcal{N}$ their maximal ideals. If $f : A \to B$ is a local homomorphism, i.e. a homomorphism such that $f(\mathcal{M}) \subset \mathcal{N}$, then*

$$\dim(B) \leq \dim(A) + \dim(B/f(\mathcal{M})B).$$

Proof Put $d = \dim(A)$ and $l = \dim(B/f(\mathcal{M})B)$. By Theorem 10.26, there exist $a_1, ..., a_d \in \mathcal{M}$ such that $(a_1, ..., a_d)$ is $\mathcal{M}$-primary and $b_1, ..., b_l \in \mathcal{N}$ such that $(\mathrm{cl}(b_1), ..., \mathrm{cl}(b_l)) \subset B/f(\mathcal{M})B$ is $\mathcal{N}/f(\mathcal{M})B$-primary.

Consequently, $(f(a_1), ..., f(a_d), b_1, ..., b_l) \subset B$ is $\mathcal{N}$-primary and $\dim(B) \leq d + l$, by Proposition 10.21. $\square$

Exercises 10.28

1. Let A be a Noetherian local ring and $a \in A$ a non-zero divisor which is not a unit. Show that $\dim(A) = \dim(A/aA) + 1$.

2. Let A be a Noetherian ring having only finitely many prime ideals. Prove that they are all maximal or minimal; in other words that $\dim(A) = 1$.

3. Let R be a Noetherian domain and $s \in R$ such that the fraction ring R_s is a field. Show that R has only finitely many non-zero prime ideals, all maximal.

10.7 Dimension of geometric rings

We can now focus our attention once again on finitely generated algebras over a field.

Example 10.29 Let K be a field and $A = K[X_1, ..., X_n]$. Then:

(i) the dimension of A is n;

(ii) all non-extendable chains of prime ideals of A have length n.

Proof If K is infinite, this is a special case of Theorem 10.18. If K is finite, let L be an algebraic closure of K. The polynomial ring $L[X_1, ..., X_n]$ is integral over its subring A. By the going-up and going-down theorems, for any non-extendable chain $\mathcal{P}_0 \subset ... \subset \mathcal{P}_l$ of prime ideals of A, there exixts a non-extendable chain $\mathcal{P}'_0 \subset ... \subset \mathcal{P}'_l$ of prime ideals of $L[X_1, ..., X_n]$ such that $\mathcal{P}'_i \cap A = \mathcal{P}_i$. Since L is infinite, we are done. $\qquad\square$

Exercise 10.30 Let $r_1, ..., r_n$ be positive integers and B a ring such that $K[X_1^{r_1}, ..., X_n^{r_n}] \subset B \subset K[X_1, ..., X_n]$. Show that all non-extendable chains of prime ideals of B have length n.

As a consequence of Corollary 10.20, we find:

Theorem 10.31 *Let A be a finitely generated algebra over an infinite field K. If Q and $\mathcal{P}$ are prime ideals of A such that $Q \subset \mathcal{P}$, then all non-extendable chains of prime ideals between Q and $\mathcal{P}$ have length*

$$\dim(A_\mathcal{P}/QA_\mathcal{P}).$$

Noetherian rings with this property are called catenary. Surprisingly, there are non-catenary Noetherian rings. More precisely, one can find a local domain R with a non-extendable chain of prime ideals having a length strictly smaller than $\dim(R)$. Fortunatly we do not care here about non-catenary Noetherian rings.

The following descriptions of the dimension need no comment.

Proposition 10.32 *Let K be an infinite field and A a finitely generated K-algebra. Then $\dim(A)$ is equal to any of the following integers:*

(i) $\max(\operatorname{trdeg}_K(A/\mathcal{P}))$, *for $\mathcal{P}$ a minimal prime ideal of A;*

(ii) $\max(\dim(A_\mathcal{M}))$, *for $\mathcal{M}$ a maximal ideal of A;*

(iii) $\max(\dim(A_{\mathcal{P}}) + \operatorname{trdeg}_K(A/\mathcal{P}))$, *for $\mathcal{P}$ a prime ideal of A.*

Definition 10.33 *Let A be a Noetherian ring. If $\dim(A/\mathcal{P}) = \dim(A)$ for all minimal prime ideals $\mathcal{P}$, we say that A is equidimensional.*

By Theorem 10.18 (ii), we get

Proposition 10.34 *Let K be a field and A a finitely generated K-algebra. Let $x_1, ..., x_n \in A$ be algebraically independent over K and such that A is integral over $K[x_1, ..., x_n]$. The following conditions are equivalent:*

(i) *the ring A is equidimensional;*

(ii) *for all minimal prime ideals $\mathcal{P}$ of A, we have $\mathcal{P} \cap K[x_1, ..., x_n] = (0)$;*

(iii) *for all prime ideals $\mathcal{P}$ of A, we have $\dim(A_{\mathcal{P}}) + \dim(A/\mathcal{P}) = n$.*

When the ring A is not equidimensional, the next result is often useful.

Proposition 10.35 *Let K be an infinite field, A a finitely generated K-algebra and $x_1, ..., x_n \in A$ elements algebraically independent over K such that A is integral over $K[x_1, ..., x_n]$.*

(i) *The function*

$$\mathcal{P} \rightarrow \dim(A_{\mathcal{P}}) + \dim(A/\mathcal{P}),$$

defined on $\operatorname{Spec}(A)$, is bounded by n and upper semi-continuous.

(ii) *If F_i is the closed set of $\operatorname{Spec}(A)$ formed by all prime ideals $\mathcal{P}$ such that*

$$\dim(A_{\mathcal{P}}) + \dim(A/\mathcal{P}) \geq i$$

then :
(a) $F_n \neq \emptyset$;
(b) there exist for all $i \leq n$ an open set $U_i \subset \operatorname{Spec}(A)$ such that $U_i \subset F_i$ and U_i is dense in F_i.

Proof We have already proved that the function is bounded by n in Theorem 10.18(i). If $\mathcal{P}$ is a prime ideal of A, let $\mathcal{Q}_m$, with $m = 1, ..., t$, be the minimal prime ideals of A contained in $\mathcal{P}$. Note that we have $\dim(A_{\mathcal{P}}) = \max_{m=1,...,t}(\dim(A/\mathcal{Q}_m)_{\mathcal{P}}))$, hence

$$\dim(A_{\mathcal{P}}) + \dim(A/\mathcal{P}) = \max_{m=1,...,t}(\dim_K(A/\mathcal{Q}_m)).$$

Fix $i \leq n$, and consider the minimal prime ideals $\mathcal{N}_j$, with $j = 1, ..., r$, of

A such that $\dim(A/\mathcal{N}_j) \geq i$. Clearly, we have

$$\dim(A_{\mathcal{P}}) + \dim(A/\mathcal{P}) \geq i \iff \mathcal{P} \in V(\bigcap_1^r \mathcal{N}_j),$$

and (i) is proved.

By Theorem 10.18(ii), if $\mathcal{P}$ is a minimal prime ideal of A such that $\mathcal{P} \cap K[x_1, ..., x_n] = (0)$, then $\dim(A/\mathcal{P}) = n$, hence $F_n \neq \emptyset$.

Next consider the minimal prime ideals $\mathcal{N}'_j$, with $j = 1, ..., s$, such that $\dim(A/\mathcal{N}'_j) < i$. Let $f \in \cap_1^s \mathcal{N}'_j$ be such that $f \notin \mathcal{N}_j$ for $j = 1, ..., r$. We claim that the open set $D(f)$ of $\text{Spec}(A)$ is contained in the closed set F_i and dense in it. Indeed, if $\mathcal{P} \in D(f)$, then $\mathcal{N}'_i \not\subset \mathcal{P}$ for $i = 1, ..., s$. Since $\mathcal{P}$ has to contain a minimal prime ideal of A, we have $\mathcal{P} \in F_i$. This shows $D(f) \subset F_i$. To conclude, note that $\mathcal{N}_j \in D(f)$ for $j = 1, ..., r$. Since $F_i = \cup_{j=1}^r V(\mathcal{N}_j)$, this shows that $D(f)$ is dense in F_i. $\square$

Proposition 10.36 *Let K be an infinite field and let A be a finitely generated equidimensional K-algebra. If $a_1, ..., a_n \in A$ and $\mathcal{P}$ is a minimal prime ideal of $(a_1, ..., a_n)$, then*

$$\dim(A/\mathcal{P}) \geq \dim(A) - n.$$

Proof Since the ring A is equidimensional, we have $\dim(A/\mathcal{P}) = \dim(A) - \dim(A_{\mathcal{P}})$. But $\dim(A_{\mathcal{P}}) \leq n$, by 10.21, and we are done. $\square$

Exercise 10.37 Let $(P_1, ..., P_r) \subset K[X_1, ..., X_n]$ be a proper ideal. Assume that P_{i+1} is not a zero divisor in $K[X_1, ..., X_n]/(P_1, ..., P_i)$ for $1 \leq i \leq r-1$. Show that $\dim(A/(P_1, ..., P_r)) = n - r$.

10.8 Exercises

1. Show that $A = \mathbb{C}[x, y, z] = \mathbb{C}[X, Y, Z]/(Z^2 - YZ + X^2)$ is a domain, that x and y are algebraically independent over $\mathbb{C}$ and that A is integral over $\mathbb{C}[x, y]$.

   Show that for all $a, b \in \mathbb{C}$ the quotient ring $A/(x-a, y-b)A$ is Artinian of length 2 and find for which $(a, b) \in \mathbb{C}^2$ this ring is not reduced.

2. Consider polynomials $f_1, ..., f_r \in \mathbb{C}[X_1, ..., X_n]$ having no common zero. Show that there exist polynomials $g_1, ..., g_r \in \mathbb{C}[X_1, ..., X_n]$ such that $\sum_{i=1}^r f_i g_i = 1$.

3. Consider $f, g \in \mathbb{C}[X_1, \ldots, X_n]$. Assume that f is irreducible and that there exists a polynomial $h \notin f\mathbb{C}[X_1, \ldots, X_n]$ such that

$$h(a_1, \ldots, a_n) \neq 0 \text{ and } f(a_1, \ldots, a_n) = 0 \Longrightarrow g(a_1, \ldots, a_n) = 0.$$

Show $g \in f\mathbb{C}[X_1, \ldots, X_n]$.

4. Let A be a domain containing $\mathbb{C}$ and B a finitely generated A-algebra. Show that there exist elements $y_1, \ldots, y_r \in B$ algebraically independent over A and a non-zero element $s \in A$ such that B_s is finite over $A_s[y_1, \ldots, y_r]$.

5. Consider $A = \mathbb{C}[X_1 \ldots, X_{n-1}]$ and $P \in A[X_n]$ a monic polynomial. Consider next a maximal ideal $\mathcal{M} = (X_1 - a_1, \ldots, X_n - a_n)/(P)$ of the ring $B = \mathbb{C}[X_1 \ldots, X_n]/(P)$ and $\mathcal{N} = \mathcal{M} \cap A$. Show that the natural homomorphism $\mathbb{C} \to B_\mathcal{M}/\mathcal{N}B_\mathcal{M}$ is an isomorphism if and only if $\frac{\partial P}{\partial X_n}(a_1, \ldots, a_n) \neq 0$.

6. Let A be a local Noetherian ring. Assume that $\dim A = n$. Show that A has infinitely many prime ideals of height i for $1 \leq i < n$.

7. Let R be a local Noetherian ring of dimension d and $\mathcal{M}$ its maximal ideal. Assume that there exists $a_1, \ldots, a_d \in \mathcal{M}$ such that $\mathcal{M} = (a_1, \ldots, a_d)$. Show that $R/(a_1, \ldots, a_i)$ is a domain for all $i \leq d$ (you can use exercise 7 (Section 5.5)).

8. Let R be a local Noetherian ring of dimension 2 and $\mathcal{M}$ its maximal ideal. Assume that there exists $a_1, a_2 \in \mathcal{M}$ such that $\mathcal{M} = (a_1, a_2)$. Show that R is a UFD.

11

Affine schemes

The decision to introduce the language of schemes at this point was not easy. An affine scheme is a topological space, equipped for each open set with a ring of algebraic functions defined on the set. Of course we have to be more careful with the definitions! We try to be so in our first two sections. From section 3 on, we relate the topological nature of an affine scheme to its algebraic structures. Although we have tried earlier in the book to prepare the reader for the language of schemes, he may have difficulties in adapting. Don't give up, there is much more commutative algebra to learn. Be sure to understand Theorems 11.15 and 11.22; they are important.

11.1   The affine space $\mathbb{A}_n$

By Corollary 10.12 of Hilbert's Nullstellensatz, there are bijective correspondences between:

- the set $\mathbb{C}^n$;

- the set $\mathrm{Spec}_m(\mathbb{C}[X_1, \ldots, X_n])$ of maximal ideals of the polynomial ring in n variables;

- the set of all $\mathbb{C}$-algebra homomorphisms $\mathbb{C}[X_1, \ldots, X_n] \to \mathbb{C}$.

To a point $x = (x_1, \ldots, x_n) \in \mathbb{C}^n$ corresponds the maximal ideal

$$\mathcal{M}_x = (X_1 - x_1, \ldots, X_n - x_n) \quad \text{of} \quad \mathbb{C}[X_1, \ldots, X_n]$$

and the evaluation homomorphism

$$e_x : \mathbb{C}[X_1, \ldots, X_n] \to \mathbb{C}, \quad \text{with} \quad e_x(P) = P(x_1, \ldots, x_n).$$

The kernel of e_x is $\mathcal{M}_x$.
    We recall that the Zariski topology on $\mathrm{Spec}(\mathbb{C}[X_1, \ldots, X_n])$ induces Zariski topologies on $\mathrm{Spec}_m(\mathbb{C}[X_1, \ldots, X_n])$ and $\mathbb{C}^n$.

To a polynomial $P \in \mathbb{C}[X_1, \ldots, X_n]$ is associated a function everywhere defined on $\mathbb{C}^n$.

To a fraction P/Q, with $P, Q \in \mathbb{C}[X_1, \ldots, X_n]$, is associated a function defined on the open set $D(Q)$ of $\mathbb{C}^n$ (open for the Zariski topology), and eventually on a larger open set if P and Q have a common factor. Note that the function P/Q is defined at the point $x \in \mathbb{C}^n$ if and only if $P/Q \in \mathbb{C}[X_1, \ldots, X_n]_{\mathcal{M}_x}$

Definition 11.1

(i)   *The fraction field $\mathbb{C}(X_1, ..., X_n)$ of $\mathbb{C}[X_1, ..., X_n]$ is the field of rational functions on $\mathbb{C}^n$.*

(ii) *The local ring $\mathbb{C}[X_1, ..., X_n]_{\mathcal{M}_x}$ is the ring of rational functions on $\mathbb{C}^n$ defined in the point x; this ring is also called the local ring of x in $\mathbb{C}^n$.*

(iii) *If $U \subset \mathbb{C}^n$ is open for the Zariski topology, $A(U) = \bigcap_{x \in U} \mathbb{C}[X_1, ..., X_n]_{\mathcal{M}_x}$ is the ring of rational functions defined in U; it is also called the ring of U in $\mathbb{C}^n$.*

(iv) *If $x \in U' \subset U$, the natural homomorphisms*

$$A(U) \rightarrow A(U') \rightarrow \mathbb{C}[X_1, ..., X_n]_{\mathcal{M}_x}$$

are the restriction homomorphisms.

Note that, by Proposition 7.10, the ring $A(\mathbb{C}^n)$ of rational functions everywhere defined is $\mathbb{C}[X_1, ..., X_n]$.

Definition 11.2

(i)   *The topological space $\mathbb{C}^n = \mathrm{Spec_m}(\mathbb{C}[X_1, \ldots, X_n])$ equipped with the function rings and the restriction homomorphisms is the complex affine space $\mathbb{A}_n$.*

(ii) *The local ring of a point x of $\mathbb{A}_n$ is denoted by $O_{\mathbb{A}_n, x}$.*

(iii) *The ring of rational functions defined on an open set U of $\mathbb{A}_n$ is denoted by $\Gamma(U, O_{\mathbb{A}_n})$.*

From now on, we shall simply say affine space rather than complex affine space.

Exercises 11.3

1. If $f \in \mathbb{C}[X_1, \ldots, X_n]$, show that $\Gamma(D(f), O_{\mathbb{A}_n}) = \mathbb{C}[X_1, \ldots, X_n]_f$.

2. Consider $f, g \in \mathbb{C}[X_1, \ldots, X_n]$, non-constant polynomials without common factors, put $U = \mathbb{A}_n \setminus V(f, g)$ and show that $\mathbb{C}[X_1, \ldots, X_n] = \Gamma(U, O_{\mathbb{A}_n})$.

11.2 Affine schemes

Let $\mathcal{I} \subset R = \mathbb{C}[X_1, ..., X_n]$ be an ideal and $X = V(\mathcal{I})$ the closed set of $\mathbb{C}^n$ consisting of all points $x \in \mathbb{C}^n$ such that $f(x) = 0$ for all $f \in \mathcal{I}$.

Putting $A = \mathbb{C}[X_1, ..., X_n]/\mathcal{I}$, we know by Corollary 10.12, that there are natural bijective correspondences between:

- the set X;

- the set $\mathrm{Spec}_m(A)$ of maximal ideals of A;

- the set of all $\mathbb{C}$-algebra homomorphisms $A \to \mathbb{C}$.

To $x = (x_1, ..., x_n) \in X$ corresponds the maximal ideal $\mathcal{M}_x = (X_1 - x_1, ..., X_n - x_n)/\mathcal{I}$ of A. Since $\mathcal{I} \subset (X_1 - x_1, ..., X_n - x_n) = \ker(e_x)$, the evaluation homomorphism $e_x : \mathbb{C}[X_1, ..., X_n] \to \mathbb{C}$ factorizes through A. This defines therefore a $\mathbb{C}$-algebra homomorphism $A \to \mathbb{C}$ whose kernel is $\mathcal{M}_x = (X_1 - x_1, ..., X_n - x_n)/\mathcal{I}$.

We identify these sets when we find it comfortable to do so.

The Zariski topology on $\mathrm{Spec}(A)$ induces once more a topology on X, via $\mathrm{Spec}_m(A)$, also called the Zariski topology. We note that X is a closed subspace of $\mathbb{C}^n$ depending only, as a topological space, on $\sqrt{\mathcal{I}}$.

In order to give to $X = \mathrm{Spec}_m(A)$ the structure of an "affine complex scheme", which will depend on $\mathcal{I}$ and not only on $\sqrt{\mathcal{I}}$, we need to define the function rings on X and the restriction homomorphisms between these rings.

Definition 11.4 *Let $x \in X$ and let $\mathcal{M}_x$ be the corresponding maximal ideal of A. The local ring $O_{X,x} = A_{\mathcal{M}_x}$ is the ring of functions on X defined in x (the local ring of x in X).*

Note that $O_{X,x} = O_{\mathbb{A}_n,x}/\mathcal{I}O_{\mathbb{A}_n,x}$, hence that there is a natural surjective homomorphism from the ring of functions on $\mathbb{A}_n$ defined in x to the ring of functions on X defined in x.

When $\mathcal{I}$ is a prime ideal, i.e. when the ring A is a domain, the fraction field $K(A)$ is the field of functions on X. For any $x \in X$, the local ring $O_{X,x}$ is a subring of $K(A)$. If U is an open subset of X, the ring of functions on X defined in U is therefore $\Gamma(U, O_X) = \bigcap_{x \in U} O_{X,x}$.

If A is not a domain, we no longer have a field containing all local rings of points. Hence if U is an open subset of X, we can no longer consider the intersection $\bigcap_{x \in U} O_{X,x}$.

The following definition is unpleasant but localizing in rings with zero divisors is always nasty.

Definition 11.5 *Let U be an open set of X. The ring $\Gamma(U, O_X)$ of functions on X defined in U is the subring of $\prod_{x \in U} O_{X,x}$ formed by the elements*

$$(a_x/b_x)_{x \in U}, \ a_x \in A, \ b_x \in A \setminus \mathcal{M}_x,$$

such that

$$b_{x'} a_x - b_x a_{x'} = 0 \quad \text{for all } x, x' \in U.$$

If $x \in U' \subset U$, the natural ring homomorphisms

$$\Gamma(U, O_X) \to \Gamma(U', O_X) \to O_{X,x}$$

are the restriction homomorphisms.

When A is a domain, $\Gamma(U, O_X)$ is naturally isomorphic to $\bigcap_{x \in U} O_{X,x}$. The proof of this fact is left to the reader.

Note that if $U \subset \mathbb{A}_n$ is an open set, there is a natural ring homomoprhism

$$\Gamma(U, O_{\mathbb{A}_n}) \to \Gamma(U \cap X, O_X).$$

Be careful, this homomorphism is not always surjective.

Next we see that the ring A is naturally isomorphic to the ring of functions defined on X.

Proposition 11.6 *The natural map*

$$A \xrightarrow{i} \Gamma(X, O_X), \quad i(a) = (a/1)_{x \in X}$$

is an isomorphism.

Proof This is a special case of Theorem 7.37. $\qquad\square$

As an obvious consequence we note that the natural homomoprhism

$$\Gamma(\mathbb{A}_n, O_{\mathbb{A}_n}) \to \Gamma(X, O_X)$$

is surjective.

CAREFUL: Let $f \in \Gamma(X, O_X)$ be a function on X everywhere defined. Consider $z \in X$ and $\mathcal{M}_z$ be the corresponding maximal ideal of $\Gamma(X, O_X)$. If $f \in \mathcal{M}_z$ then $f(z) = 0$. If f is nilpotent, then $f(x) = 0$ for all $x \in X$. Consequently, if the ring $\Gamma(X, O_X)$ is not reduced, there exists non-zero functions $g \in \Gamma(X, O_X)$ such that $g(x) = 0$ for all $x \in X$. The existence of such functions is certainly one of the difficulties of the language of schemes.

Definition 11.7

(i) *The topological space $X = \mathrm{Spec_m}(A)$ equipped with the function rings and the restriction homomorphisms is a complex affine scheme.*

(ii) *The affine scheme X is a closed subscheme of $\mathbb{A}_n$.*

(iii) *The ring $A = \Gamma(X, O_X)$ is the ring of the affine scheme X.*

(iv) *If the ring A is reduced, then X is a reduced affine scheme.*

(v) *If the ring A is a domain, then X is an affine variety. In this case the fraction field of A is denoted by $K(X)$ and called the field of functions on X.*

From now on we shall say affine scheme for complex affine scheme.

CAREFUL: If $\mathcal{I}$ is an ideal of $\mathbb{C}[X_1, ..., X_n]$, the subsets $V(\mathcal{I}^2)$ and $V(\mathcal{I})$ of $\mathbb{C}^n$ are clearly bijective. Furthermore the topological spaces $V(\mathcal{I}^2)$ and $V(\mathcal{I})$ are homeomorphic. But the affine schemes $V(\mathcal{I}^2)$ and $V(\mathcal{I})$ are distinct when $\mathcal{I} \neq \mathcal{I}^2$.

If $X = V(\mathcal{I}) \subset \mathbb{A}_n$, then the ideal $\mathcal{I}$ of $\mathbb{C}[X_1, ..., X_n]$ is the kernel of the natural ring homomorphism $\mathbb{C}[X_1, ..., X_n] = \Gamma(\mathbb{A}_n, O_{\mathbb{A}_n}) \to \Gamma(V(\mathcal{I}), O_{V(\mathcal{I})})$. As a consequence, we get:

Proposition 11.8 *The correspondence $\mathcal{I} \leftrightarrow V(\mathcal{I})$ between closed subschemes of $\mathbb{A}_n$ and ideals of $\mathbb{C}[X_1, \ldots, X_n]$ is bijective.*

11.3 Closed and open subschemes of an affine scheme

First, we study the open sets of an affine scheme $X = \mathrm{Spec_m}(A)$ defined by a function.

If $f \in A$, then $D(f) = \{x \in X \text{ such that } f(x) \neq 0\}$ is an open set in X. Indeed $X \setminus D(f)$ is the closed set $V_\mathrm{m}(fA) \subset \mathrm{Spec_m}(A)$.

Exercise 11.9 Show that $\Gamma(D(f), O_X) = A_f$.

Proposition 11.10

(i) *An open set of X is a finite union of open sets of the form $D(f)$.*

(ii) *$D(f) \cap D(g) = D(fg)$.*

Proof Let $\mathcal{I}$ be an ideal of A and $F = V_\mathrm{m}(\mathcal{I})$. If $\mathcal{I} = (f_1, \ldots, f_r)$, it is clear that $U = X \setminus F = \bigcup_1^r D(f_i)$ or equivalently that $x \in U$ if and only if $f_i(x) \neq 0$ for some i. This proves (i), and (ii) is obvious. □

Proposition 11.11 *If $U \subset X$ is an open set and if $g \in \Gamma(U, O_X)$, the set*

$$D(g) = \{x \in U \quad \text{such that} \quad g(x) \neq 0\}$$

is an open subset of U.

Proof Put $U = \bigcup_1^r D(f_i)$ and consider for each i the restriction homomorphism

$$\pi_i : \Gamma(U, O_X) \to \Gamma(D(f_i), O_X) = A_{f_i}.$$

If $x \in D(f_i)$, then $g(x) \neq 0$ if and only if $\pi_i(g)(x) \neq 0$. Consequently, $D(g) \cap D(f_i)$ is open. $\qquad\square$

Exercise 11.12 If $f, g \in A$ are such that f is not a zero divisor and that g is not a zero divisor modulo f, put $U = X \setminus V(f, g)$ and show that the natural homomorphism $A \to \Gamma(U, O_X)$ is an isomorphism.

Now, we introduce the closed subschemes of an affine scheme.

Definition 11.13 *Let $X = \mathrm{Spec}_m(A)$ be an affine scheme.*

(i) *If $\mathcal{J}$ is an ideal of A, the affine scheme $\mathrm{Spec}_m(A/\mathcal{J})$ is the closed subscheme $V(\mathcal{J})$ of X.*

(ii) *If $\mathcal{P}$ is a prime ideal of A, then $\mathrm{Spec}_m(A/\mathcal{P})$ is the closed subvariety $V(\mathcal{P})$ of X. If we denote this variety by X', the ring $A_{\mathcal{P}}$ is the local ring $O_{X,X'}$ of X' in X.*

We note that a closed subscheme of an affine scheme is an affine scheme.

Examples 11.14 Consider $x \in X = \mathrm{Spec}_m(A)$ and $\mathcal{M}_x \in \mathrm{Spec}_m(A)$ the corresponding maximal ideal of the ring $A = \Gamma(X, O_X)$.

1. The subvariety $\{x\} = V(\mathcal{M}_x)$ of X is concentrated in x. The ring of functions defined on this subvariety is $A/\mathcal{M}_x \simeq \mathbb{C}$. The local ring of the subvariety $\{x\}$ in X is $A_{\mathcal{M}_x} = O_{X,x}$, i.e. the local ring of the point x of X.

2. If $\mathcal{M}_x \neq \mathcal{M}_x^2$, the subscheme $V(\mathcal{M}_x^2)$ is concentrated in x, but is not a subvariety.

Theorem 11.15 *Let $X = \mathrm{Spec}_m(A)$ be an affine scheme.*

(i) *The correspondence $F \longleftrightarrow V_m(\bigcap_{x \in F} \mathcal{M}_x)$, between closed subsets of the topological space X and reduced closed subschemes of X, is bijective.*

(ii) *This correspondence induces a bijective correspondence between irreducible closed subsets of X and prime ideals of A.*

Proof (i) Let F be closed in X. There exists an ideal $\mathcal{I} \subset A$ such that $x \in F \Longleftrightarrow \mathcal{I} \subset \mathcal{M}_x$. But this is equivalent to $\sqrt{\mathcal{I}} \subset \mathcal{M}_x$. Since A is a Jacobson ring, this shows that $\sqrt{\mathcal{I}} = \bigcap_{x \in F} \mathcal{M}_x$. This proves that F and $\sqrt{\mathcal{I}}$ do characterize each other, hence the correspondence is bijective.

(ii) First assume that $\sqrt{\mathcal{I}}$ is a prime ideal. If $F = F_1 \cup F_2$, put $\mathcal{I}_i = \bigcap_{x \in F_1} \mathcal{M}_x$. Then

$$\sqrt{\mathcal{I}} = \Big(\bigcap_{x \in F_1} \mathcal{M}_x \Big) \bigcap \Big(\bigcap_{x \in F_2} \mathcal{M}_x \Big) = \mathcal{I}_1 \bigcap \mathcal{I}_2.$$

Since a prime ideal is irreducible there exists i such that $\sqrt{\mathcal{I}} = \mathcal{I}_i$. This implies

$$x \in F \Longleftrightarrow \sqrt{\mathcal{I}} \subset \mathcal{M}_x \Longleftrightarrow \mathcal{I}_i \subset \mathcal{M}_x \Longleftrightarrow x \in F_i$$

and F is irreducible.

Next assume that the radical ideal $\sqrt{\mathcal{I}}$ is not prime. Consider a minimal primary decomposition $\sqrt{\mathcal{I}} = \bigcap_1^r \mathcal{P}_j$, with $r > 1$. Let $X_i = V_{\mathrm{m}}(\mathcal{P}_i)$ and F_i be the closed subset of X carried by X_i. Clearly, we have $F = \bigcup_1^r F_i$.

Now, we claim that $F_i \not\subset \bigcup_{j \neq i} F_j$. Note that $\mathcal{P}_i = \bigcap_{x \in F_i} \mathcal{M}_x$. Then, $F_i \subset \bigcup_{j \neq i} F_j$ implies

$$\sqrt{\mathcal{I}} = \bigcap_{x \in F} \mathcal{M}_x = \bigcap_{j \neq i} \Big(\bigcap_{x \in F_j} \mathcal{M}_x \Big) = \bigcap_{j \neq i} \mathcal{P}_j.$$

This is a contradiction, since we had a minimal primary decomposition of $\sqrt{\mathcal{I}}$. This proves $F_i \neq F$ for all i, hence F is reducible. □

Note, since this is important, the essential meaning of Theorem 11.15: A reduced closed subscheme of an affine scheme is characterized by its underlying topological space.

Corollary 11.16 *Let $X = \mathrm{Spec}_{\mathrm{m}}(A)$ be an affine scheme. Then X is irreducible as a topological space if and only if the nilradical of A is a prime ideal.*

Proof Since $\bigcap_{x \in X} \mathcal{M}_x = \sqrt{(0)}$, this is a consequence of Theorem 11.15. □

Definition 11.17 *Let $X = \mathrm{Spec}_{\mathrm{m}}(A)$ be an affine scheme. If $Y = \mathrm{Spec}_{\mathrm{m}}(A/\mathcal{I})$ and $Z = \mathrm{Spec}_{\mathrm{m}}(A/\mathcal{J})$ are closed subschemes of X, then:*

(i) *their intersection $Y \cap Z$ is the closed subscheme $\mathrm{Spec}_{\mathrm{m}}(A/(\mathcal{I} + \mathcal{J}))$ of X;*

(ii) *their union $Y \cup Z$ is the closed subscheme $\mathrm{Spec}_{\mathrm{m}}(A/(\mathcal{I} \cap \mathcal{J}))$ of X.*

We recall the title of this chapter: Affine schemes. We noted that the closed subchemes of an affine scheme are affine schemes, hence naturally studied here. In order to be able to present and use the "affine open subschemes" of an affine scheme, we find it convenient to introduce its "open subschemes", although they are not all affine.

Definition 11.18 *Let X be an affine scheme and $U \subset X$ an open set.*

(i) *If $x \in U$, the ring of functions on U defined in x is $O_{U,x} = O_{X,x}$.*

(ii) *If $U' \subset U$ is an open set, the ring of functions on U defined in U' is*
$$\Gamma(U', O_U) = \Gamma(U', O_X)$$

(iii) *The topological space U equipped with its function rings and the restriction homomorphisms is an open subscheme of the affine scheme X.*

(iv) *An open subscheme of an affine scheme is a quasi-affine scheme.*

Now we can present, as announced, the affine open subschemes of an affine scheme.

Consider an open subset $U \subset X$ and put $B = \Gamma(U, O_X)$. For $x \in U$, the kernel $\mathcal{N}_x$ of the composition homomorphism

$$B = \Gamma(U, O_X) \to O_{X,x} \to O_{X,x}/\mathcal{M}_x O_{X,x} \simeq \mathbb{C}$$

is a maximal ideal of B. In other words, the inverse image $\mathcal{N}_x$ of $\mathcal{M}_x O_{X,x}$ by the restriction homomorphism $B = \Gamma(U, O_X) \to O_{X,x}$ is a maximal ideal of B. Note that this induces, for each $x \in U$, a natural local homomorphism $B_{\mathcal{N}_x} \to O_{X,x} = O_{U,x}$.

If B is a finitely generated $\mathbb{C}$-algebra, then $\mathrm{Spec}_{\mathrm{m}}(B)$ is an affine scheme and we have defined a map $U \to \mathrm{Spec}_{\mathrm{m}}(B)$, associating to $x \in U$ the maximal ideal $\mathcal{N}_x \in \mathrm{Spec}_{\mathrm{m}}(B)$. We claim furthermore that this map is continuous. Indeed, $g \in B \setminus \mathcal{N}_x$ if and only if $g(x) \neq 0$.

Definition 11.19 *Let $U \subset X$ be an open set. We say that U is an affine open subscheme of X if the following conditions are satisfied:*

(i) *the ring $B = \Gamma(U, O_X)$ is a finitely generated $\mathbb{C}$-algebra;*

(ii) *the map $U \to \mathrm{Spec}_{\mathrm{m}}(B)$, $x \to \mathcal{N}_x$ is a homeomorphism;*

(iii) *for all $x \in U$ the natural homomorphism $B_{\mathcal{N}_x} \to O_{U,x}$ is an isomorphism.*

Note that since the map $U \to \mathrm{Spec}_{\mathrm{m}}(B)$ is continuous, our condition (ii) is satisfied if and only if this map is bijective.

Proposition 11.20 *Let X be an affine scheme and $A = \Gamma(X, O_X)$. If $f \in A$, then $D(f)$ is an affine open subscheme of X whose ring of functions is $A_f \simeq A[Z]/(Zf - 1)$.*

Proof We have seen in Exercise 11.9 that $A_f = \Gamma(D(f), O_X)$. We claim that the map

$$D(f) \to \operatorname{Spec}_m(A_f)$$

is bijective. Indeed, this map is induced by the natural bijective correspondence between the prime ideals of A not containing f and the prime ideals of A_f. More precisely, if $x \in D(f)$, then $\mathcal{M}_x A_f$ is the image of x by this map. So we must prove that if $\mathcal{N}$ a maximal ideal of A_f, then the unique prime ideal $\mathcal{P}$ of A such that $\mathcal{P} A_f = \mathcal{N}$ is maximal. We recall that $\mathcal{P} = \mathcal{N} \cap A$. Since A_f is a finitely generated $\mathbb{C}$-algebra, $A_f/\mathcal{N} \simeq \mathbb{C}$. This implies $A/(\mathcal{N} \cap A) \simeq \mathbb{C}$, hence $\mathcal{P}$ is a maximal ideal of A. □

Exercise 11.21 If $f, g \in A$ are elements such that f is not a zero divisor and that g is not a zero divisor modulo f, show that $X \setminus V(f, g)$ is not an affine open subscheme of X.

11.4 Functions defined on an open set

If U is an open set of an affine scheme X, there exist finitely many functions f_i everywhere defined on X, such that $U = \bigcup_i D(f_i)$. We use such a decomposition of U to describe the ring of functions defined on the open set U of X.

Theorem 11.22 *Let U be an open set in an affine scheme $X = \operatorname{Spec}_m(A)$. Let $f_1, ..., f_n \in A$ be functions such that $U = \bigcup D(f_i)$. If $l_{ij} : A_{f_i} \to A_{f_i f_j} = A_{f_j f_i}$ are the localization homomorphisms, there is a natural isomorphism*

$$\Gamma(U, O_X) \simeq \ker[h : \oplus A_{f_i} \to \oplus_{i<j} A_{f_i f_j}],$$

with

$$h(a_1, ..., a_n) = (l_{ij}(a_i) - l_{ji}(a_j))_{i<j}.$$

This theorem is a special case of the next proposition.

Proposition 11.23 *Let U be an open set in an affine scheme $X = \operatorname{Spec}_m(A)$. Let $f_1, ..., f_n \in A$ be functions such that $U = \bigcup D(f_i)$. If M is a finitely generated A-module, there is a natural isomorphism between the following A-modules:*

(i)

$$K = \ker[h : \oplus_i M_{f_i} \to \oplus_{i<j} M_{f_i f_j}], \quad h(a_1, ..., a_n) = (l_{ij}(a_i) - l_{ji}(a_j))_{i<j},$$

where $l_{ij} : M_{f_i} \to M_{f_i f_j} = M_{f_j f_i}$ *are the localization homomorphisms;*

(ii) *the submodule* M' *of* $\prod_{x \in U} M_{\mathcal{M}_x}$ *formed by the elements*

$$(a_{\mathcal{M}_x}/s_{\mathcal{M}_x})_{x \in U}, \quad \text{with } a_{\mathcal{M}_x} \in M, \text{ and } s_{\mathcal{M}_x} \in A \setminus \mathcal{M}_x$$

such that $s_{\mathcal{M}_x} a_{\mathcal{M}_{x'}} - s_{\mathcal{M}_{x'}} a_{\mathcal{M}_x} = 0$ *for all* $x, x' \in U.$

The kernel of the natural homomorphism $M \to M' \simeq K$ *is*

$$\bigcup_{l \geq 0} ((0) : (f_1, ..., f_n)^l).$$

This homomorphism is not injective if and only if there exists $\mathcal{P} \in \mathrm{Ass}(M)$ *such that* $(f_1, ..., f_n) \subset \mathcal{P}$ *and is zero if and only if* $U \cap \mathrm{Supp}(M) = \emptyset.$

Proof First, we describe a natural homomorphism $\pi : K \to M'$. Consider an element $a = (a_1/f_1^{l_1}, ..., a_n/f_n^{l_n}) \in K$ with $a_i \in M$ for all i. Then

$$a = (a_1 f_1^r / f_1^{r+l_1}, ..., a_n f_n^r / f_n^{r+l_n}), \quad \text{for} \quad r \geq 0.$$

But, $l_{ij}(a_i/f_i^{l_i}) = l_{ji}(a_j/f_j^{l_j}) \in M_{f_i f_j}$ implies

$$a_i f_i^r f_j^{r+l_j} = a_j f_j^r f_j^{r+l_i} \in M \quad \text{for} \quad r >> 0.$$

Consequently, if $x \in U$ and i is such that $f_i \notin \mathcal{M}_x$, then $a_i f_i^r / f_i^{r+l_i} \in M_{\mathcal{M}_x}$ only depends on $x \in K$, and not on i. This defines our homomorphism $\pi : K \to M'$.

Next, we define an inverse homomorphism. Since the open set $D(f_i) \subset U$, there is an obvious homomorphism $M' \to \prod_{x \in D(f_i)} M_{\mathcal{M}_x}$ whose image is contained in the submodule formed by the elements

$$(a_{\mathcal{M}_x}/s_{\mathcal{M}_x})_{x \in D(f_i)}, \quad \text{with } a_{\mathcal{M}_x} \in M, \text{ and } s_{\mathcal{M}_x} \in A \setminus \mathcal{M}_x$$

such that $s_{\mathcal{M}_x} a_{\mathcal{M}_{x'}} - s_{\mathcal{M}_{x'}} a_{\mathcal{M}_x} = 0$ for all $x, x' \in D(f_i).$

Now $D(f_i)$ is the affine scheme $\mathrm{Spec}_m(A_{f_i})$. By Theorem 7.37, there is a natural isomorphism between M_{f_i} and the submodule of $\prod_{x \in D(f_i)} (M)_{\mathcal{M}_x}$ formed by the elements

$$(a_{\mathcal{M}_x}/s_{\mathcal{M}_x})_{x \in D(f_i)}, \quad \text{with } a_{\mathcal{M}_x} \in M_{f_i} \text{ and } s_{\mathcal{M}_x} \in A_{f_i} \setminus \mathcal{M}_x A_{f_i}$$

such that $s_{\mathcal{M}_x} a_{\mathcal{M}_{x'}} - s_{\mathcal{M}_{x'}} a_{\mathcal{M}_x} = 0$ for all $x, x' \in D(f_i).$

Hence we have found, for each i, a natural homomorphism $M' \to M_{f_i}$. This defines a homomorphism $M' \to \oplus_i M_{f_i}$. Checking that its image is in K is straightforward. Proving that it is the inverse of π is fastidious. Do it only if you feel like it!

Consider now the natural homomorphism

$$\phi : M \to \oplus_i M_{f_i}, \quad x \to (x/1, ..., x/1).$$

Obviously its image is in K and just as obviously $(x/1, ..., x/1) = (0, ..., 0)$ if and only if $f_i^n x = 0$ for all i and for $n >> 0$, in other words if and only if $x(f_1, ..., f_n)^l = (0)$ for $l >> 0$.

This shows that ϕ is not injective if and only if there exists $\mathcal{P} \in \mathrm{Ass}(M)$ such that $(f_1, ..., f_n) \subset \mathcal{P}$. Furthermore $\phi = 0$ if and only if $(f_1, ..., f_n)^l M = (0)$ for $l >> 0$, but this is equivalent to $U \cap \mathrm{Supp}(M) = \emptyset$. $\square$

Remark 11.24 *When we know enough about depth and the conditions $\mathcal{S}_i$, we shall see that the natural homomorphism $M \to K$ is bijective if and only if $\mathrm{depth}(M_{\mathcal{P}}) \geq 2$ for all $\mathcal{P} \in \mathrm{Supp}(M)$ such that $(f_1, \ldots, f_n) \subset \mathcal{P}$.*

11.5 Dimension of an affine scheme

Definition 11.25 *The dimension $\dim(X)$ of an affine scheme X is the dimension of the ring $\Gamma(X, O_X)$ of functions everywhere defined on X.*

Note that $\dim(X)$ is the largest integer n such that there exists a strict chain of subvarieties of X:

$$V_0 \subset V_1 \subset ... \subset V_n.$$

Theorem 11.26 *Let X be an affine variety and K the field of functions on X. For all subvarieties $X' \subset X$, we have*

$$\dim(O_{X,X'}) + \dim(X') = \dim(X) = \mathrm{trdeg}_{\mathbb{C}}(K).$$

In particular, for all $x \in X$, we have

$$\dim(O_{X,x}) = \dim(X) = \mathrm{trdeg}_{\mathbb{C}}(K).$$

This is Theorem 10.20.

As a consequence of Proposition 10.35, we get

Theorem 11.27 *Let X be an affine scheme.*

(i) *The function $x \to \dim(O_{X,x})$ is upper semi-continuous.*

(ii) *If $F \subset X$ is the closed set such that $\dim(O_{X,x}) = dim(X) \iff x \in F$, there exists a open subset $U \subset X$ such that $U \subset F$ and U is dense in F.*

11.6   Irreducible components of an affine scheme

Consider an affine scheme $X = \text{Spec}_{\text{m}}(A)$. In the ring A, let

$$(0) = (\cap_1^s \mathcal{I}_i) \cap (\cap_1^t \mathcal{J}_j)$$

be a minimal primary decomposition organized such that

- the prime ideals $\mathcal{P}_i = \sqrt{\mathcal{I}_i}$, with $1 \leq i \leq s$, are the minimal prime ideals of A;

- the prime ideals $\mathcal{P}'_j = \sqrt{\mathcal{J}_j}$, with $1 \leq j \leq t$, are the embedded associated prime ideals of A.

Since A is a Jacobson ring, the nilradical $\sqrt{(0)} = \cap_1^s \mathcal{P}_i$ of A is equal to $\cap_{x \in X} \mathcal{M}_x$.

Let F_i be the closed set of X defined by the ideal $\mathcal{P}_i$, i.e. $x \in F_i \Longleftrightarrow \mathcal{P}_i \subset \mathcal{M}_x$, We know, by Theorem 11.15, that $F_1, \ldots F_s$ are the irreducible components of the topological space X. Note that F_i is the topological space carrying the scheme $V_{\text{m}}(\mathcal{I}_i)$.

Definition 11.28

(i) *The closed subschemes $X_i = \text{Spec}_{\text{m}}(A/\mathcal{I}_i)$ are the irreducible components of X.*

(ii) *For each non-minimal associated prime ideal $\mathcal{P}'_j$, we say that X has an imbedded component along the subvariety $\text{Spec}_{\text{m}}(A/\mathcal{P}'_j)$ (we recall that $\mathcal{J}_t$ is not uniquely defined).*

(iii) *The closed subscheme $\text{Spec}_{\text{m}}(A/\cap_1^s \mathcal{P}_i)$ of X is the reduced scheme X_{red} of X.*

(iv) *If X has only one irreducible component, we say that the affine scheme X is irreducible.*

(v) *If X is connected as a topological space, we say that the affine scheme X is connected.*

Note that an affine variety is an irreducible reduced affine scheme. Note also that an irreducible scheme is obviously connected.

Definition 11.29 *If all the irreducible components of an affine scheme X have the same dimension, we say that X is equidimensional.*

Proposition 11.30 *An affine scheme X is equidimensional if and only if for all $x \in X$, we have $\dim(O_{X,x}) = \dim(X)$.*

Proof Let $\mathcal{P}_i$, with $i = 1, ..., n$, be the minimal prime ideals of the ring $\Gamma(X, O_X)$. Consider the affine varieties $X_i = \operatorname{Spec}_m(A/\mathcal{P}_i)$. We have

$$\dim(O_{X,x}) = \max_{x \in X_i}(\dim(O_{X_i,x})).$$

But $\dim(O_{X_i,x}) = \dim(X_i)$ for all $x \in X_i$, by Theorem 11.26.

If X is equidimensional, then $\dim(X_i) = \dim(X)$ for all i and $\dim(O_{X,x}) = \dim(X)$ for all $x \in X$.

If X is not equidimensional, there exists i such that $\dim(X_i) < \dim(X)$. Since $X_i \not\subset \bigcup_{j \neq i} X_j$, there exists $x \in X_i$ such that $x \notin X_j$ for $j \neq i$. This implies $\dim(O_{X,x}) = \dim(X_i) < \dim(X)$. $\qquad\square$

Proposition 11.31 *Let $X = \operatorname{Spec}_m(A)$ be an affine scheme without imbedded component. If there exists a dense open set $U \subset X$ such that $O_{X,x}$ is a domain for all $x \in U$, then X is a variety.*

Proof Let $(0) = (\cap_{s=1}^n \mathcal{I}_s)$ be the minimal primary decomposition of (0) in the ring A. Note that the non-empty open set $X \setminus V(\cap_{s \neq j} \mathcal{I}_s)$ is contained in the closed set $V(\mathcal{I}_j)$. This shows $V(\mathcal{I}_j) \cap U \neq \emptyset$ for all j.

Consider $x \in U$. Since $O_{X,x}$ is a domain, there is a unique integer s such that $x \in V(\mathcal{I}_s)$. Hence the closed sets $V(\mathcal{I}_s) \cap U$ of U are pairwise disjoint. Since their union U is connected, there is only one such closed set and X is irreducible. Hence $\sqrt{(0)}$ is a prime ideal $\mathcal{P}$ of A. Assume there exists a non-zero element $a \in \mathcal{P}$. Since a is nilpotent, $a/1 \in O_{X,x} = A_{\mathcal{M}_x}$ is zero for all $x \in U$. In other words, $((0) : a) \not\subset \mathcal{M}_x$ for all $x \in U$. This shows

$$V_m((0) : a) \cap U = \emptyset.$$

But $((0) : a) \subset \mathcal{P}$, since $\mathcal{P}$ is the only prime ideal associated to A. This implies

$$V_m(\mathcal{P}) \cap U = \emptyset,$$

an obvious contradiction. $\qquad\square$

11.7   Exercises

1. Consider

$$z_1 = (0,0), \quad z_2 = (0,1), \quad z_3 = (1,0),$$

$$z_4 = (1,1) \in \mathbb{A}_2 = \operatorname{Spec}_m(\mathbb{C}[X_1, X_2]).$$

Give a system of generators of the radical ideal formed by all polynomials $f \in \mathbb{C}[X_1, X_2]$ such that $f(z_i) = 0$ for all i.

2. Show that for all $0 \leq i \leq d$, there exists a polynomial $f \in \mathbb{C}[X_1, X_2]$ of degree d and such that $\mathbb{C}[X_1, X_2]/(X_1, f)$ is an Artinian ring of length i.

3. Let $f_1, \ldots, f_r \in \mathbb{C}[X_1, \ldots, X_n]$ be non-zero elements such that

$$\mathrm{cl}(f_{i+1}) \in \mathbb{C}[X_1, \ldots, X_n]/(f_1, \ldots, f_i)$$

is not a zero divisor for $1 \leq i$. Show that if

$$\mathrm{Spec_m}(\mathbb{C}[X_1, \ldots, X_n]/(f_1, \ldots, f_r))$$

is not empty, then it is an equidimensional affine scheme of dimension $(n - r)$.

4. Let $X = \mathrm{Spec_m}(A)$ be an affine variety. If $\mathcal{P} \subset A$ is a prime ideal such that $\mathrm{ht}(\mathcal{P}) \geq 2$, show that the open subscheme $X \setminus V(\mathcal{P})$ of X is not affine.

5. Let Z be an affine closed subvariety of $\mathbb{A}_n = \mathrm{Spec_m}(\mathbb{C}[X_1, \ldots, X_n])$. Show that there exists $f \in \mathbb{C}[X_1, \ldots, X_n]$ such that $Z = V(f)$ if and only if $\dim X = n - 1$.

6. Consider the ring homomorphism

$$\pi : \mathbb{C}[X_1, \ldots, X_4] \to \mathbb{C}[U^d, U^{d-1}V, UV^{d-1}, V^d]$$

where

$$\pi(X_1) = U^d, \ \pi(X_2) = U^{d-1}V, \ \pi(X_3) = UV^{d-1}, \ \text{and} \ \pi(X_4) = V^d.$$

Put $\mathcal{P} = \ker \pi$. Show that $V(\mathcal{P})$ is a closed subvariety of dimension 2 of $\mathbb{A}_4$ and that $\mathcal{P}$ cannot be generated by less than d elements. Hint: Show that $\mathcal{P}$ is generated by one homogeneous polynomial of degree 2 and $(d-1)$ homogeneous polynomials of degree $(d-1)$.

7. If $H \in \mathbb{C}[X_1, \ldots, X_n]$ has degree one, we say that $V(H)$ is a hyperplane of $\mathbb{A}_n$. If two hyperplanes of $\mathbb{A}_4$ are distinct and intersect, their intersection is called an affine plane of $\mathbb{A}_4$.

Show that the variety

$$X = V(X_1X_4 - X_2X_3) \subset \mathbb{A}_4 = \mathrm{Spec_m}(\mathbb{C}[X_1, \ldots, X_4])$$

contains infinitely many affine planes of $\mathbb{A}_4$.

8. Show that the variety $X = V(\sum_{i=1}^4 X_i^3) \subset \mathbb{A}_4 = \mathrm{Spec_m}(\mathbb{C}[X_1, \ldots, X_4])$ contains 27 affine planes of $\mathbb{A}_4$.

12

Morphisms of affine schemes

An affine scheme is a topological space equipped with its rings of functions. We recall that these functions are algebraic. To a "morphism of affine schemes" $X \to Y$ should correspond a map $\Gamma(Y, O_Y) \to \Gamma(X, O_X)$ of rings of functions. Now, we want this map to be a $\mathbb{C}$-algebra homomorphism. Conversely, we shall see that a $\mathbb{C}$-algebra homomorphism $\Gamma(Y, O_Y) \to \Gamma(X, O_X)$ induces a continuous map of topological spaces $X \to Y$. We note, and this is important, that if this map is a homeomorphism, the ring homomorphism need not be an isomorphism. In our second section, we are careful in studying open and closed immersions of affine schemes. Then, after having defined the product of two affine schemes, we show, in the fifth section, an important result: the product of two varieties is a variety. Consequently, we can prove a first theorem on intersection (Corollary 12.28): if X and Y are subvarieties of $\mathbb{A}_n$, an irreducible component of $X \cap Y$ has dimension at least $\dim(X) + \dim(Y) - n$.

12.1 Morphisms of affine schemes

Consider r polynomials $G_1, ..., G_r \in \mathbb{C}[X_1, ..., X_n]$. They induce a map

$$\pi : \mathbb{C}^n \to \mathbb{C}^r, \quad \pi(x_1, ..., x_n) = (G_1(x_1, ..., x_n), ..., G_r(x_1, ..., x_n)).$$

They induce as well a homomorphism of polynomial rings

$$f : \mathbb{C}[Y_1, ..., Y_r] \to \mathbb{C}[X_1, ..., X_n], \quad \text{with} \quad f(Y_i) = G_i.$$

Let $\mathcal{M}_x = (X_1 - x_1, ..., X_n - x_n) \subset \mathbb{C}[X_1, ..., X_n]$ be the maximal ideal corresponding to the point $x \in \mathbb{A}_n$, and $\mathcal{N}_{\pi(x)} = (Y_1 - G_1(x_1, ..., x_n), ..., Y_r - G_r(x_1, ..., x_n)) \subset \mathbb{C}[Y_1, ..., Y_r]$ the maximal ideal corresponding to the point $\pi(x) \in \mathbb{A}_r$. Then, it is easy to check that

$$f^{-1}(\mathcal{M}_x) = \mathcal{N}_{\pi(x)}.$$

In other words, recalling that $\mathbb{C}^l$ is the underlying set of $\mathrm{Spec}_{\mathrm{m}}(\mathbb{C}[X_1, ..., X_l])$, the map π is induced by the map

$$\mathrm{Spec}_{\mathrm{m}}(\mathbb{C}[X_1, ..., X_n]) \to \mathrm{Spec}_{\mathrm{m}}(\mathbb{C}[Y_1, ..., Y_r]), \quad \mathcal{M} \to f^{-1}(\mathcal{M}).$$

This map is a morphism from $\mathbb{A}_n$ to $\mathbb{A}_r$.

More generally, let $X = \mathrm{Spec_m}(B)$ and $Y = \mathrm{Spec_m}(A)$ be affine schemes. We claim that a $\mathbb{C}$-algebra homomorphism $f : A \to B$ defines a continuous map $\pi_f : X \to Y$.

Consider $x \in X$ and $\mathcal{M}_x$ the corresponding maximal ideal of B. The natural injective homomorphisms of finitely generated $\mathbb{C}$-algebras

$$\mathbb{C} \hookrightarrow A/f^{-1}(\mathcal{M}_x) \hookrightarrow B/\mathcal{M}_x = \mathbb{C}$$

show that $f^{-1}(\mathcal{M}_x)$ is a maximal ideal of A. If $y \in Y = \mathrm{Spec_m}(B)$ is the corresponding point, put $y = \pi_f(x)$.

Now if $a \in A$, we have $\pi_f^{-1}(D(a) \cap \pi_f(X)) = D(f(a)) \subset X$. This proves that π_f is continuous. We note that $f(\mathcal{M}_{f(x)}) \subset \mathcal{M}_x$ for all $x \in X$ and f induces therefore, for all $x \in X$, a local homomorphism of local rings

$$f_x : O_{Y, \pi_f(x)} \to O_{X, x}.$$

Definition 12.1 *The map $\pi_f : X \to Y$ defined by f is a morphism of affine schemes. If f is an isomorphism, π_f is an isomorphism.*

Consider $Z = \mathrm{Spec_m}(C)$ another affine scheme and $\pi_g : Z \to X$ a morphism of affine schemes defined by a $\mathbb{C}$-algebra homomorphism $g : B \to C$. Then the maps $\pi_{g \circ f}$ and $\pi_f \circ \pi_g$ are obviously identical, hence the composition of two composable morphisms of affine schemes is a morphism of affine schemes.

Examples 12.2

1. For any affine scheme X, the natural homomorphism $\mathbb{C} \to \Gamma(X, O_X)$ defines a (structural) morphism $X \to \mathbb{A}_0$.

2. Let $f : \mathbb{C}[Y_1, ..., Y_r] \to \mathbb{C}[X_1, ..., X_n]$ be a ring homomorphism. If $G_i = f(Y_i)$, then $\pi_f(x_1, ..., x_n) = (G_1(x_1, ..., x_n), ..., G_r(x_1, ..., x_n))$.

3. If $G_1, ..., G_r$ are polynomials of degree one such that $G_1, ..., G_r$ and 1 are linearly independent, π_f is the projection from $\mathbb{A}_n$ on $\mathbb{A}_r$ with centre $V(G_1, ..., G_r) \subset \mathbb{A}_n$. We note that $x \in V(G_1, ..., G_r) \Leftrightarrow \pi_f(x) = (0, ..., 0)$.

4. if $G_i = X_i$ for $1 \le i \le r$, then $\pi_f(x_1, ..., x_n) = (x_1, ..., x_r)$.

5. Let $A = \mathbb{C}[y_1, ..., y_r]$ be a quotient ring of $\mathbb{C}[Y_1, ..., Y_r]$. Then $\mathrm{Spec_m}(A)$ is a closed subscheme of $\mathbb{A}_r$. We say that the morphism $\mathrm{Spec_m}(A) \to \mathbb{A}_r$ defined by the natural surjective homomorphism $\mathbb{C}[Y_1, ..., Y_r] \to A$ is a closed embedding of $\mathrm{Spec_m}(A)$ in $\mathbb{A}_r$ and we write $\mathrm{Spec_m}(A) \subset \mathbb{A}_r$.

6. More generally, let A be a finitely generated $\mathbb{C}$-algebra and $\mathcal{I}$ an ideal of A. The natural surjective homomorphism $A \to A/\mathcal{I}$ induces a morphism $\mathrm{Spec_m}(A/\mathcal{I}) \to \mathrm{Spec_m m}(A)$. This is the closed imbedding

$$\mathrm{Spec_m}(A/\mathcal{I}) \subset \mathrm{Spec_m}(A).$$

7. Consider $X = \mathrm{Spec_m}(\mathbb{C}[x_1, ..., x_n])$ an affine closed subscheme of $\mathbb{A}_n$. Let $H_i \in \mathbb{C}[X_1, ..., X_n]$, for $i = 1, ..., m$, be polynomials of degree one such that $H_1, ..., H_m$ and 1 are linearly independent. The natural homomorphism $f : \mathbb{C}[H_1, ..., H_m] \to \mathbb{C}[x_1, ..., x_n]$ induces a morphism

$$\pi_f : X \to \mathbb{A}_m, \quad \text{with} \quad \pi_f(x_1, ..., x_n) = (H_1(x_1, ..., x_n), ..., H_m(x_1, ..., x_n)).$$

This is the projection of X to the affine space $\mathbb{A}_m = \mathrm{Spec_m}(\mathbb{C}[H_1, ..., H_m])$. It is obviously the composition of the imbedding $X \subset \mathbb{A}_n$ and the projection, with centre $V(H_1, ..., H_m)$, of $\mathbb{A}_n$ on $\mathbb{A}_m$.

Proposition 12.3 *A morphism $\pi : X \to Y$ of affine schemes is an isomorphism if and only if it is a homeomorphism of topological spaces and for each $x \in X$ the local homomorphism $O_{Y,\pi(x)} \to O_{X,x}$ is an isomorphism.*

Proof Let $f : \Gamma(Y, O_Y) \to \Gamma(X, O_X)$ be the ring homomorphism defining π. If f is an isomorphism, the two conditions are obviously satisfied. Conversely, assume they are satified: then the local isomorphisms, induced by f, induce an isomorphism

$$\prod_{y \in Y} O_{Y,y} = \prod_{x \in X} O_{Y,\pi(x)} \simeq \prod_{x \in X} O_{X,x}.$$

Then, the following commutative diagram

$$\begin{array}{ccc} \Gamma(Y, O_Y) & \xrightarrow{f} & \Gamma(X, O_X) \\ \cap & & \cap \\ \prod\limits_{y \in Y} O_{Y,y} & \simeq & \prod\limits_{x \in X} O_{X,x} \end{array}$$

and Proposition 11.6 show that f is an isomorphism. $\qquad\qquad\square$

12.2 Immersions of affine schemes

Definition 12.4 *Let $\pi : X \to Y$ be a morphism of affine schemes. If $\pi(X)$ is an affine open subscheme U of Y and if $\pi : X \to U$ an isomorphism of affine schemes, we say that π is an open immersion.*

Example 12.5 Let $g \in \Gamma(Y, O_Y) = A$ be a function defined on Y. Consider the affine scheme $X = \mathrm{Spec_m}(A[Z]/(gZ - 1))$. The natural homomorphism $f : A \to A[Z]/(gZ - 1)$ defines an open immersion $\pi_f : X \to Y$ whose image is the open affine scheme $D(g)$.

Proposition 12.6 *A morphism* $\pi : X \to Y$ *of affine schemes is an open immersion if and only if the following conditions are satisfied:*

(i) *the subset* $\pi(X)$ *of* Y *is open and homeomorphic to* X;

(ii) *for each* $x \in X$ *the local homomorphism* $O_{Y,\pi(x)} \to O_{X,x}$ *is an isomorphism.*

Proof If π is an open immersion the conditions are obviously satisfied.

Conversely, by Proposition 12.3, it suffices to show that the open set $U = \pi(X)$ is an affine subscheme of Y. The local isomorphisms induce an isomorphism $\Gamma(U, O_Y) \simeq \Gamma(X, O_X)$. This shows that $B = \Gamma(U, O_Y)$ is a finitely generated $\mathbb{C}$-algebra. The map $\mathrm{Spec_m}(\Gamma(X, O_X)) \to U$ is a homeomorphism. The natural map $\phi : U \to \mathrm{Spec_m}(B)$ is obvioulsly the inverse homeomorphism. Next consider $y \in U$. If $x = \pi^{-1}(y)$ and $\mathcal{N}_y = \phi(y)$, the factorizations

$$B_{\mathcal{N}_y} \to O_{Y,y} \simeq O_{X,x} \simeq B_{\mathcal{N}_y}$$

show that the natural local homomorphism $B_{\mathcal{N}_y} \to O_{Y,y}$ is an isomorphism. Hence U is an affine open subscheme of Y.   □

Proposition 12.7 *If* $\pi : X \to Y$ *is a morphism of affine schemes and* U *an affine subscheme of* Y, *then* $\pi^{-1}(U)$ *is an affine open subscheme of* X. *There is a natural isomorphism of* $\mathbb{C}$-*algebras*

$$\Gamma(\pi^{-1}(U), O_X) \simeq \Gamma(U, O_U) \otimes_{\Gamma(Y, O_Y)} \Gamma(X, O_X).$$

Proof The rings $\Gamma(U, O_U)$ and $\Gamma(X, O_X)$ are finitely generated $\mathbb{C}$-algebras. Put $A = \Gamma(Y, O_Y)$. By the natural homomorphisms

$$A \to \Gamma(U, O_U) \quad \text{and} \quad A \to \Gamma(X, O_X)$$

they both have the structure of an A-algebra. Then $B = \Gamma(U, O_U) \otimes_A \Gamma(X, O_X)$ is a finitely generated $\mathbb{C}$-algebra. Consider the affine scheme $Z = \mathrm{Spec_m}(B)$. The $\mathbb{C}$-algebra homomorphism

$$g : \Gamma(X, O_X) \to B \quad g(a) = 1 \otimes a$$

defines a morphism $\psi_g : Z \to X$. We can check at once that ψ_g is an open immersion whose image is $\pi^{-1}(U)$, i.e. that Z is homeomorphic to $\pi^{-1}(U)$ and that the local homomorphisms $O_{X,\psi_g(z)} \to O_{Z,z}$ are isomorphisms for all $z \in Z$.   □

Corollary 12.8 *If U and V are affine open subschemes of an affine scheme Y, then $U \cap V$ is an affine open subscheme of Y such that*

$$\Gamma(U \cap V, O_X) \simeq \Gamma(U, O_U) \otimes_{\Gamma(Y, O_Y)} \Gamma(V, O_V).$$

Proof Apply Proposition 12.7 to the morphism of affine schemes $U \to Y$ (or to $V \to Y$ if you prefer!). □

Definition 12.9 *Let $\pi : X \to Y$ be a morphism of affine schemes. If there exists a closed scheme $Y' \subset Y$ such that π factorizes through an isomorphism $X \to Y'$, we say that π is a closed immersion.*

Note that π is a closed immersion if and only if the corresponding ring homomorphism $f : \Gamma(Y, O_Y) \to \Gamma(X, O_X)$ is surjective.

Example 12.10 Let B be a quotient ring of $A = \Gamma(Y, O_Y)$. If $Z = \mathrm{Spec_m}(B)$ the closed embedding $Z \subset Y$ defines (thank God!) a closed immersion of Z in Y. In particular, if $\mathcal{M}_y$ is the maximal ideal of A corresponding to a point $y \in Y$, then the surjective homomorphism $A \to A/\mathcal{M}_y$ defines the closed immersion of $\{y\}$ in Y.

Proposition 12.11 *A morphism $\pi : X \to Y$ of affine schemes is a closed immersion if and only if the following conditions are satisfied:*

(i) *the subset $\pi(X)$ of Y is closed and homeomorphic to X;*

(ii) *for each $x \in X$ the local homomorphism $O_{Y,\pi(x)} \to O_{X,x}$ is surjective.*

Proof If π is a closed immersion the conditions are obviously satisfied.

Conversely, put $B = \Gamma(X, O_X)$ and $A = \Gamma(Y, O_Y)$ and consider the homomorphism $f : A \to B$ defining π. We want to prove that f is surjective. Consider the A-module $C = \mathrm{coker}(f)$. For $y \in Y$, let $\mathcal{M}_y$ be the corresponding maximal ideal of A and $S = A - \mathcal{M}_y$. We want to show that for all $y \in Y$, we have $S^{-1}C = C_{\mathcal{M}_y} = (0)$.

If $y \notin \pi(X)$, there exists $s \in S$ such that $D(s) \cap \pi(X) = \emptyset$. Consequently, $D(f(s)) = \emptyset \subset \mathrm{Spec_m}(B)$. In other words, $f(s)$ is contained in all maximal ideals of B. Since B is a Jacobson ring, $f(s)$ is nilpotent and $B_{f(s)} = (0)$. Consequently, $C_s = (0)$, hence $S^{-1}C = C_{\mathcal{M}_y} = (0)$.

If $y = \pi(x)$, let $\mathcal{N}_x$ be the maximal ideal of B corresponding to x. We have $f^{-1}(\mathcal{N}_x) = \mathcal{M}_y$ and the homomorphism $A_{\mathcal{M}_y} \to B_{\mathcal{N}_x}$ is surjective. Hence to prove that $S^{-1}C = C_{\mathcal{M}_y} = (0)$, we need only to show that $f(S)^{-1}B = B_{\mathcal{N}_x}$, in other words that all prime ideals of B disjoint from $f(S)$ are contained in $\mathcal{N}_x$. Let $\mathcal{P}$ be such a prime ideal. Consider the variety $X' = V(\mathcal{P}) \subset X$ and the closed set $Y' = \pi(X') \subset Y$ homeomorphic to X'. But $f^{-1}(\mathcal{P}) \cap S = \emptyset$ implies $f^{-1}(\mathcal{P}) \subset \mathcal{M}_y$, hence $y \in Y'$. Consequently, $\pi^{-1}(y) = x \in X'$ and $\mathcal{P} \subset \mathcal{N}_x$. $\qquad\qquad\qquad\qquad\qquad\qquad\qquad\qquad\qquad\qquad\qquad\qquad\qquad\quad\square$

Example 12.12 The natural homomorphism $\mathbb{C}[X]/(X^n) \to \mathbb{C}[X]/(X)$ induces a homomorphism $\mathbb{A}_0 \to \mathrm{Spec}_m(\mathbb{C}[X]/(X^n))$ which is not an isomorphism for $n > 1$, but a closed immersion.

12.3 Local description of a morphism

Theorem 12.13 *Let X and Y be affine schemes. Assume that there exist finitely many affine open subschemes V_i of X covering X and morphisms $\pi_i : V_i \to Y$ such that $\pi_i \mid_{V_i \cap V_j} = \pi_j \mid_{V_i \cap V_j}$ for all (i,j). Then there exists a morphism $\pi : X \to Y$ such that $\pi \mid V_i = \pi_i$ for all i.*

Note that for all (i,j), the open set $V_i \cap V_j$ is an affine scheme, hence that $\pi_i \mid_{V_i \cap V_j} : V_i \cap V_j \to Y$ is well-defined.

Proof of 12.13 Since an open set is a union of open sets of the form $D(g)$, we can assume that there exist $g_i \in B = \Gamma(X, O_X)$ such that $V_i = D(g_i)$. The morphisms π_i are induced by homomorphisms $h_i : A = \Gamma(Y, O_Y) \to B_{g_i}$.

Consider the localization homomorphisms $l_{ij} : B_{g_i} \to B_{g_i g_j}$. Since $\pi_i \mid_{V_i \cap V_j} = \pi_j \mid_{V_i \cap V_j}$, we have $l_{ij} \circ h_i = l_{ji} \circ h_j$. In other words the homomorphism

$$h : A \to \oplus_i B_{g_i}, \quad h(a) = (h_i(a))_i,$$

has its image in the kernel of

$$\oplus_i B_{g_i} \to \oplus_{i,j} B_{g_i g_j}.$$

By Theorem 11.22 this kernel is B. It is easy to chek that the morphism $\pi : X \to Y$ corresponding to h satisfies our needs. $\qquad\qquad\qquad\qquad\square$

Definition 12.14 *Let U and V be quasi-affine schemes. A morphism $\pi : V \to U$ is the following data:*

(i) *an affine covering $(U_i)_i$ of U;*

(ii) *an affine covering $(V_i)_i$ of V;*

(iii) *morphisms $\pi_i : V_i \to U_i$ such that $\pi_i \mid_{V_i \cap V_j} = \pi_j \mid_{V_i \cap V_j}$ for all i,j.*

Note that if $\pi : X \to Y$ is a morphism of affine schemes and $U \subset X$ an open subscheme, then $\pi \mid_U \colon U \to Y$ is a morphism from the quasi-affine scheme U to the affine scheme Y.

We can now define immersions of quasi-affine schemes.

Definition 12.15 *A morphism $\pi : X \to Y$ from a quasi-affine scheme X to an affine scheme Y is an open immersion if $\pi(X)$ is an open set of Y homeomorphic to X and for each $x \in X$ the local homomorphism $O_{Y,\pi(x)} \to O_{X,x}$ is an isomorphism.*

Definition 12.16 *A morphism $\pi : X \to Y$ of quasi-affine schemes is a closed immersion if $\pi(X)$ is a closed set of Y homeomorphic to X and for each $x \in X$ the local homomorphism $O_{Y,\pi(x)} \to O_{X,x}$ is surjective.*

Definition 12.17 *A closed and open immersion $\pi : X \to Y$ of quasi-affine schemes is an isomorphism.*

Exercise 12.18 Let R be a finitely generated $\mathbb{C}$-algebra and $\mathcal{I}$ an ideal of R. Consider the R-algebras $A = \oplus_{n \geq 0} \mathcal{I}^n T^n \subset B = R[T]$ and the natural morphisms of affine schemes

$$\mathrm{Spec}_m(B) \xrightarrow{\pi} \mathrm{Spec}_m(A) \xrightarrow{\psi} \mathrm{Spec}_m(R).$$

Note that if $t \in \mathcal{I}$, we have $A_t = B_t$. Show that if $U = \mathrm{Spec}(R) - V_m(\mathcal{I})$ the induced morphism of quasi-affine schemes

$$\pi \mid_{\pi^{-1}(\psi^{-1}(U))} \colon \pi^{-1}(\psi^{-1}(U)) \simeq \psi^{-1}(U)$$

is an isomorphism.

12.4   Product of affine schemes

Definition 12.19 *Let X and Y be affine schemes. The product $X \times_{\mathbb{C}} Y$ of X and Y is the affine scheme whose ring of functions is*

$$\Gamma(X \times_{\mathbb{C}} Y, 0_{X \times_{\mathbb{C}} Y}) = \Gamma(X, O_X) \otimes_{\mathbb{C}} \Gamma(Y, O_Y).$$

The projections $p_1 : X \times_{\mathbb{C}} Y \to X$ and $p_2 : X \times_{\mathbb{C}} Y \to Y$ are the morphisms associated to the natural $\mathbb{C}$-algebras homomorphisms

$$\Gamma(X, O_X) \to \Gamma(X, O_X) \otimes_{\mathbb{C}} \Gamma(Y, O_Y), \quad a \to a \otimes 1$$

and

$$\Gamma(Y, O_Y) \to \Gamma(X, O_X) \otimes_{\mathbb{C}} \Gamma(Y, O_Y), \quad a \to 1 \otimes a.$$

It is easy to verify that the set carrying $X \times_{\mathbb{C}} Y$ is $X \times Y$. One needs to work a bit more to show that $X \times_{\mathbb{C}} Y$ is indeed the solution of a universal problem. In other words, to show that given an affine scheme Z with morphisms $q_1 : Z \to X$ and $q_2 : Z \to Y$, there is a unique morphism $\pi : Z \to X \times_{\mathbb{C}} Y$ such that $q_i = p_i \circ \pi$ for $i = 1, 2$.

Exercise 12.20 Show that $\mathbb{A}_{n+m} \simeq \mathbb{A}_n \times_{\mathbb{C}} \mathbb{A}_m$.

Definition 12.21 *Let* $\pi : X \to Z$ *and* $\psi : Y \to Z$ *be morphisms of affine schemes.*

(i) *The product* $X \times_Z Y$ *of* X *and* Y *is the affine scheme whose ring of functions is*

$$\Gamma(X \times_Z Y, 0_{X \times_Z Y}) = \Gamma(X, O_X) \otimes_{\Gamma(Z, O_Z)} \Gamma(Y, O_Y).$$

(ii) *The projections* $p_1 : X \times_Z Y \to X$ *and* $p_2 : X \times_Z Y \to Y$ *are the morphisms associated to the natural* $\mathbb{C}$-*algebra homomorphisms*

$$\Gamma(X, O_X) \to \Gamma(X, O_X) \otimes_{\Gamma(Z, O_Z)} \Gamma(Y, O_Y), \quad a \to a \otimes 1$$

and

$$\Gamma(Y, O_Y) \to \Gamma(X, O_X) \otimes_{\Gamma(Z, O_Z)} \Gamma(Y, O_Y), \quad a \to 1 \otimes a.$$

Once more the reader will have to check by himself that this product is, as it should be, a solution of the universal problem. In other words, that if W is an affine scheme and $q_1 : W \to X$ and $q_2 : W \to Y$ are morphisms such that $\pi \circ q_1 = \psi \circ q_2$, then there exists a unique morphism $\tau : W \to X \times_Z Y$ such that $q_i = p_i \circ \tau$ for $i = 1, 2$.

Exercises 12.22

1. Consider a morphism of affine schemes $\pi : X \to Z$ and the identity automorphism of Z. Show that the projection morphism $p_1 : X \times_Z Z \to X$ is an isomorphism

2. If U and V are affine open subschemes of Y, use Corollary 12.8 to show that
$$U \cap V = U \times_Y V.$$

Note that the natural surjective ring homomorphism

$$\Gamma(X, O_X) \otimes_{\mathbb{C}} \Gamma(Y, O_Y) \to \Gamma(X, O_X) \otimes_{\Gamma(Z, O_Z)} \Gamma(Y, O_Y)$$

induces a closed immersion $X \times_Z Y \to X \times_{\mathbb{C}} Y$.

Definition 12.23 *If $\pi : X \to Z$ is a morphism of affine schemes, the closed subscheme $X \times_Z Z$ of $X \times_{\mathbb{C}} Z$ is the graph of π.*

Proposition 12.24 *Let $\pi : X \to Y$ be a morphism of affine schemes. If Z is a closed subscheme of Y, the projection morphism $X \times_Y Z \to X$ is a closed immersion.*

Proof From the definition of the product we see that $X \times_Y Z$ is a closed subscheme of $X \times_Y Y$. Since, by Exercise 12.22, $p_1 : X \times_Y Y \to X$ is an isomorphism, we are done. $\qquad\square$

Definition 12.25 *Let $\pi : X \to Y$ be a morphism of affine schemes.*

(i) *If Z is a closed (resp. open) affine subscheme of Y, the closed (open) subscheme $p_1(X \times_Y Z)$ of X is the inverse image $\pi^{-1}(Z)$ of Z in X.*

(ii) *If y is a point of Y, the inverse image $\pi^{-1}(y)$ of the reduced subscheme concentrated in y is the fibre of $y \in Y$ by π.*

In other words, if $\mathcal{M}_y \subset \Gamma(Y, \mathcal{O}_Y)$ is the maximal ideal corresponding to the point $y \in Y$ and if $f : \Gamma(Y, \mathcal{O}_Y) \to \Gamma(X, \mathcal{O}_X)$ is the homomorphism defining π, then $f(\mathcal{M}_y)\Gamma(X, \mathcal{O}_X)$ is the ideal of the fibre of y in the ring of X.

Examples 12.26 Consider the morphisms $\pi_i : \mathbb{A}_1 \to \mathbb{A}_2$ defined by the $\mathbb{C}$-algebra homomorphisms $f_i : \mathbb{C}[X_1, X_2] \to \mathbb{C}[T]$:

$$f_1(X_i) = T^i; \quad f_2(X_1) = T(T-1), \ f_2(X_2) = T(T-1)(T-2); \quad f_3(X_i) = T^{i+1}.$$

1. The morphism π_1 is a closed immersion.

2. The fibre $\pi_2^{-1}(0,0)$ is a closed reduced subscheme of $\mathbb{A}_1$ whose ring is $\mathbb{C}[T]/T(T-1)$. This fibre contains two points.

3. The fibre $\pi_3^{-1}(0,0)$ is a closed irreducible, non-reduced, subscheme of $\mathbb{A}_1$ whose ring is $\mathbb{C}[T]/T^2$. It contains a unique point.

12.5 Dimension, product and intersection

Theorem 12.27 *Let X and Y be affine varieties.*

(i) *$X \times_{\mathbb{C}} Y$ is an affine variety.*

(ii) *$\dim(X \times_{\mathbb{C}} Y) = \dim(X) + \dim(Y)$.*

Proof We first prove that the affine scheme $Z = X \times_{\mathbb{C}} Y$ is irreducible.

Assume $Z = Z_1 \cup Z_2$, where Z_i is a closed subscheme of Z. If $x \in X$, the fibre $p_1^{-1}(x)$ is isomorphic to Y, hence is a subvariety of Z. Since $p_1^{-1}(x) = (p_1^{-1}(x) \cap Z_1) \cup (p_1^{-1}(x) \cap Z_2)$, there exists i such that $p_1^{-1}(x) \subset Z_i$. Consequently, if $X_i = \{x \in X, \ p_1^{-1}(x) \subset Z_i\}$, we have $X = X_1 \cup X_2$.

We claim that X_i is closed. To see this, consider, for all $y \in Y$, the closed subscheme of X

$$X_i(y) = p_1(Z_i \cap p_2^{-1}(y)) = \{x \in X, \ (x, y) \in Z_i\}.$$

It is clear that $X_i = \cap_{y \in Y} X_i(y)$. This shows that X_i is closed. Since X is irreducible, there exists i such that $X = X_i$. Obviously, this implies $Z = Z_i$.

Next we prove that $Z = X \times_{\mathbb{C}} Y$ is reduced. Consider $h = \sum_i f_i \otimes_{\mathbb{C}} g_i$, with $f_i \in \Gamma(X, O_X)$ and $g_i \in \Gamma(Y, O_Y)$, a function on Z. Obviously, we can assume that the functions g_i are linearly independent on $\mathbb{C}$. If h is nilpotent, then

$$\sum_i f_i(x) \otimes_{\mathbb{C}} g_i(y) = 0 \quad \text{for all } (x, y) \in Z.$$

Fix $x \in X$. Since $\sum_i f_i(x) \otimes_{\mathbb{C}} g_i(y) = 0$ for all $y \in Y$, the function $\sum_i f_i(x) \otimes_{\mathbb{C}} g_i \in \Gamma(Y, O_Y)$ is contained in all maximal ideals of the ring $\Gamma(Y, O_Y)$. But this ring is a Jacobson domain, hence $\sum_i f_i(x) \otimes_{\mathbb{C}} g_i = 0$. Since the functions g_i are linearly independent, this means $f_i(x) = 0$ for all i. We have therefore proved $f_i(x) = 0$ for all $x \in X$. Since the ring $\Gamma(X, O_X)$ is a Jacobson domain, this shows $f_i = 0 \in \Gamma(X, O_X)$, hence $h = 0$, and (i) is proved.

Now (ii) is a straightforward consequence of the normalization lemma. Indeed, consider $T_1, ..., T_d \in \Gamma(X, O_X)$ (resp. $S_1, ..., S_l \in \Gamma(Y, O_Y)$) algebraically independent elements such that $\Gamma(X, O_X)$ (resp. $\Gamma(Y, O_Y)$) is integral over $\mathbb{C}[T_1, ..., T_d]$ (resp. $\mathbb{C}[S_1, ..., S_l]$).

It is clear that the polynomial ring, in $d + l$ variables,

$$\mathbb{C}[T_1, ..., T_d, S_1, ..., S_l] = \mathbb{C}[T_1, ..., T_d] \otimes_{\mathbb{C}} \mathbb{C}[S_1, ..., S_l]$$

is a subring of the function ring

$$\Gamma(X \times_{\mathbb{C}} Y, 0_{X \times_{\mathbb{C}} Y}) = \Gamma(X, O_X) \otimes_{\mathbb{C}} \Gamma(Y, O_Y).$$

We claim that $\Gamma(X \times_{\mathbb{C}} Y, 0_{X \times_{\mathbb{C}} Y})$ is integral over $\mathbb{C}[T_1, ..., T_d, S_1, ..., S_l]$. Indeed, it is generated, as an algebra, by elements of the forms $f \otimes_{\mathbb{C}} 1$, with $f \in \Gamma(X, O_X)$, and $1 \otimes_{\mathbb{C}} g$, with $g \in \Gamma(Y, O_Y)$. Now, since $f \in \Gamma(X, O_X)$ is integral over $\mathbb{C}[T_1, ..., T_d]$, an element $f \otimes_{\mathbb{C}} 1$ is integral over $\mathbb{C}[T_1, ..., T_d, S_1, ..., S_l]$. So is an element $1 \otimes_{\mathbb{C}} g$ for the same reason, and we are done. $\square$

Corollary 12.28 *Let X and Y be closed subvarieties of $\mathbb{A}_n$. If Z is an irreducible component of $X \cap Y$, then*

$$\dim(Z) + n \geq \dim(X) + \dim(Y).$$

Proof Consider $R = \mathbb{C}[X_1, ..., X_n]$, the ring of $\mathbb{A}_n$. Let $\mathcal{P} = (f_1, ..., f_l)$ and $\mathcal{Q} = (g_1, ..., g_m)$ be the prime ideals of X and Y in R. If $A = \Gamma(X \cap Y, O_{X \cap Y})$, there is a natural isomorphism

$$A \simeq \mathbb{C}[X_1, ..., X_n, T_1, ..., T_n]$$
$$/((f_i(X_1, ..., X_n))_{1 \leq i \leq l}, (g_k(T_1, ..., T_n))_{1 \leq k \leq m}, (X_j - T_j)_{1 \leq j \leq n}).$$

By Proposition 10.36, it is sufficient to show that the ring

$$B = \mathbb{C}[X_1, ..., X_n, T_1, ..., T_n]/((f_i(X_1, ..., X_n))_{1 \leq i \leq l}, (g_k(T_1, ..., T_n))_{1 \leq k \leq m})$$

is equidimensional of dimension $\dim(X) + \dim(Y)$.

But there is an obvious isomorphism

$$B \simeq \mathbb{C}[X_1, ..., X_n]/(((f_i(X_1, ..., X_n))_{1 \leq i \leq l})$$
$$\otimes_{\mathbb{C}} \mathbb{C}[T_1, ..., T_n]/(((g_k(T_1, ..., T_n))_{1 \leq k \leq m}).$$

Consequently, the ring

$$B \simeq \Gamma(X, O_X) \otimes_{\mathbb{C}} \Gamma(Y, O_Y) = \Gamma(X \times_{\mathbb{C}} Y, 0_{X \times_{\mathbb{C}} Y})$$

is a domain of dimension $\dim(X) + \dim(Y)$ by Theorem 12.27. □

Definition 12.29 *Let X and Y be closed subvarieties of $\mathbb{A}_n$. If $\dim(Z) + n = \dim(X) + \dim(Y)$ for each irreducible component Z of $X \cap Y$, we say that X and Y intersect properly in $\mathbb{A}_n$.*

12.6 Dimension and fibres

Proposition 12.30 *Let $\pi : X \to Y$ be a morphism of affine schemes. For $x \in X$, we have*

$$\dim(O_{Y, \pi(x)}) + \dim(O_{\pi^{-1}(\pi(x)), x}) \geq \dim(O_{X, x}).$$

If furthermore the ring homomorphism $\Gamma(Y, O_Y) \to \Gamma(X, O_X)$ is injective, there exists a non-empty open subset $U \subset X$ such that

$$\dim(O_{Y, \pi(x)}) + \dim(O_{\pi^{-1}(\pi(x)), x}) = \dim(O_{X, x}).$$

for all $x \in U$.

Proof The inequality is a special case of Corollary 10.27.

If $A = \Gamma(Y, O_Y)$ and $B = \Gamma(X, O_X)$, assume now that the ring homomorphism $A \to B$, defining π is injective.

By the normalization lemma, there exist $y_1, ..., y_n \in A$ algebraically independent over $\mathbb{C}$ and such that A is finite over $R = \mathbb{C}[y_1, ..., y_n]$. Put $S = R - 0$ and note that $S^{-1}B$ is finitely generated as an $S^{-1}R$-algebra. By the normalization lemma, yet again, there exist $z_1, ..., z_m \in S^{-1}B$ algebraically independent over the field $S^{-1}R$ and such that $S^{-1}B$ is finite over $S^{-1}R[z_1, ..., z_m]$. We note first that we can assume $z_i \in B$, for all i, by getting rid of the denominators if necessary. Next we recall that B is a $\mathbb{C}$-algebra of finite type, say $B = \mathbb{C}[x_1, ..., x_m]$. Obviously, there exists $s \in S$ such that x_i is integral over $R_s[z_1, ..., z_m]$ for all i. Consequently B_s is finite over $R_s[z_1, ..., z_m]$.

Let $\mathcal{P}$ be a minimal prime ideal of B such that $\mathcal{P}B_s \cap R_s[z_1, ..., z_m] = (0)$. If $\mathcal{M} \subset B$ is a maximal ideal such that $s \notin \mathcal{M}$ and $\mathcal{P} \subset \mathcal{M}$, we claim that

$$\dim(B_\mathcal{M}) = \dim(B_\mathcal{M}/\mathcal{P}B_\mathcal{M}) = n + m.$$

Indeed $\mathcal{N} = \mathcal{M} \cap R[z_1, ..., z_m]$ is a maximal ideal. Hence it contains a chain of prime ideals, of length $n + m$. It induces a chain of prime ideals of $R_s[z_1, ..., z_m]$ contained in $\mathcal{N}R_s[z_1, ..., z_m]$. By the going-down theorem, there is a chain of prime ideals of $B_s/\mathcal{P}B_s$, contained in $\mathcal{M}B_s/\mathcal{P}B_s$ and lying over this chain. This shows $\dim(B_\mathcal{M}/\mathcal{P}B_\mathcal{M}) = n + m$.

Now let $\mathcal{Q}_i$, with $i = 1, ..., r$, be the minimal prime ideals of B_s such that $\mathcal{Q}_i \cap R_s[z_1, ..., z_m] \neq (0)$. Since $\mathcal{Q}_i \not\subset \mathcal{P}$ for all i, there exists $t \in \bigcap_{1 \leq i \leq r} \mathcal{Q}_i$ such that $t \notin \mathcal{P}$. The open set $U = D(st)$ is not empty and we have proved that it satisfies the assertion. $\qquad\square$

Exercise 12.31 Let Y be an affine variety and $A = \Gamma(Y, O_Y)$. Consider B, a finitely generated A-algebra and $x_1, \dots, x_n \in B$ such that $B = A[x_1, \dots, x_n]$. Put $X = \operatorname{Spec_m}(B)$ and show that all fibres of the morphism $\pi : X \to Y$ have dimension less than or equal to n. Show furthermore that if all fibres of π have dimension n, then the elements $x_1, \dots, x_n$ are algebraically independent over A.

12.7   Finite morphisms

Definition 12.32 *A morphism $\pi : X \to Y$ of affine schemes is finite if the corresponding ring homomorphism $\Gamma(Y, O_Y) \to \Gamma(X, O_X)$ is finite.*

Note that all fibres of a finite morphism are finite. Indeed, let $\mathcal{M}$ be a maximal ideal of $A = \Gamma(Y, O_Y)$. Since the ring $B = \Gamma(X, O_X)$ is finite over A, it is clear that $B/\mathcal{M}B$ is finite over the field $A/\mathcal{M}$. Consequently $B/\mathcal{M}B$ is Artinian and has only a finite number of maximal ideals.

Theorem 12.33 *If* $\pi : X \to Y$ *is a finite morphism of affine schemes,* $\pi(X)$
is closed in Y.

Proof Let $\mathcal{I}$ be the kernel of $f : \Gamma(Y, O_Y) \to \Gamma(X, O_X)$. If $\mathcal{M}$ is a maximal
ideal of $\Gamma(Y, O_Y)$ containing $\mathcal{I}$, we know by Theorem 8.26 that there exists a
maximal ideal $\mathcal{N}$ of $\Gamma(X, O_X)$ such that $f^{-1}(\mathcal{N}) = \mathcal{M}$. This proves $\pi(X) = V_{\mathfrak{m}}(\mathcal{I})$. $\qquad\qquad\square$

The normalization lemma can be restated as follows:

Theorem 12.34 *If* $X \subset \mathbb{A}_n$ *is a closed affine subscheme of dimension* d,
there exists a projection $\mathbb{A}_n \to \mathbb{A}_d$ *such that the induced projection morphism*
$X \to \mathbb{A}_d$ *is finite and surjective.*

Exercise 12.35 Let X be an affine variety of dimension n and $X \to \mathbb{A}_n$
a finite morphism. If $\mathbb{A}_n \to \mathbb{A}_r$ is a projection, show that all fibres of the
composition morphism $X \to \mathbb{A}_r$ have dimension $n - r$.

12.8   Exercises

1. Let X and Y be varieties and $\pi : X \to Y$ a morphism. Assume that
   $\dim X = 1$ and that $\pi(X)$ is not a point. Show that all the fibers of π
   are finite.

2. Put $Z = \operatorname{Spec}_{\mathfrak{m}}(\mathbb{C}[X, Y, T]/(XT - Y))$. Then,

   $$\mathbb{C}[X, Y] \subset \mathbb{C}[X, Y, T]/(XT - Y)$$

   induces a morphism $\pi : Z \to \mathbb{A}_2$.

   Show that π induces an isomorphism of open affine schemes $\pi^{-1}(D(X)) \simeq D(X)$, hence an open immersion $\pi^{-1}(D(X)) \subset \mathbb{A}_2$. Prove that the fiber
   $\pi^{-1}(0, 0)$ is isomorphic to $\mathbb{A}_1$. Finally, check that $\pi(Z) = D(X) \cup \{(0, 0)\}$.

Definition 12.36 *A morphism of varieties* $\pi_f : X \to Y$ *is called rational if the corresponding homomorphism of domains* $f : \Gamma(Y, O_Y) \to \Gamma(X, O_X)$ *is injective.*

A rational morphism of varieties $\pi_f : X \to Y$ *is called birational if the injective homomorphism* $K(Y) \to K(X)$ *is an isomorphism.*

3. Let $\pi_f : X \to Y$ be a birational morphism of affine varieties. Show that
   there exists an affine open subscheme $D(s) \subset Y$ such that π_f induces an
   isomorphism $\pi_f^{-1}(D(s)) \simeq D(s)$.

4. Let $\pi_f : X \to Y$ be a rational morphism of varieties. Assume that the field extension $K(Y) \subset K(X)$ is algebraic. Show that there exists an affine open subscheme $D(s) \subset Y$ such that the induced morphism of affine schemes $\pi_f^{-1}(D(s)) \to D(s)$ is finite.

5. In the preceding exercise, show that we can choose s such that for all $y \in D(s)$, the Artinian ring $\Gamma(\pi_f^{-1}(y), O_{\pi_f^{-1}(y)})$ has length $[K(X) : K(Y)]$.

6. In the preceding exercise, show that we can choose s such that for all $y \in D(s)$, the fibre $\pi_f^{-1}(y)$ contains exactly $[K(X) : K(Y)]$ distinct points. To do so, consider $b \in \Gamma(X, O_X)$ such that $K(X) = K(Y)[b]$ and $P(Y, T) \in K(Y)[T]$ a minimal polynomial of b over $K(Y)$. Then choose t such that the coefficients of P are functions defined on $D(t)$ and that for each $y \in D(t)$ the polynomial $P(y, T) \in \mathbb{C}[T]$ has no multiple root. Finally take the intersection of $D(t)$ with the open set found in the preceding exercise.

7. Consider an affine variety $Y = \mathrm{Spec_m}(A)$ where A is an integrally closed finitely generated $\mathbb{C}$-algebra. Consider a finite extension of domains $A[T_1, \ldots, T_n] \subset B$. Show that each fibre of the induced morphism $\pi : \mathrm{Spec_m}(B) \to Y$ is equidimensional of dimension n.

8. Let $\pi_f : X \to Y$ be a rational morphism of varieties. Show that there exists an affine open subscheme $D(s) \subset Y$ such that the ring

$$\Gamma(\pi_f^{-1}(D(s)), O_X)$$

is finite over a polynomial ring over the ring $\Gamma(D(s), O_Y)$. Use the preceding exercise to show that there exists an affine open set $U \subset Y$ and an integer n such that for $y \in U$ the fibre $\pi^{-1}(y)$ is equidimensional of dimension n.

13

Zariski's main theorem

If V is an affine variety, a non-empty open set U of V is dense in V. Hence a point $x \in V$ is a connected component of V if and only if $V = \{x\}$, in other words if and only if $\dim(V) = 0$. Consequently, if X is an affine scheme and $x \in X$, then $\{x\}$ is a connected component of X if and only if it is an irreducible component. In other words if and only if the maximal ideal $\mathcal{M}_x$ of $\Gamma(X, O_X)$ is also a minimal prime ideal of $\Gamma(X, O_X)$. We note that if $X = \mathrm{Spec}_m(A)$ is a finite set, then the ring A is Artinian and each point of X forms a connected component of X.

Next consider $\pi : X \to Y$ a morphism of affine schemes, $x \in X$ and $\pi^{-1}(\pi(x))$ the fibre of $\pi(x)$. Then x is a connected component of $\pi^{-1}(\pi(x))$ if and only if $\dim(O_{\pi^{-1}(\pi(x)),x}) = 0$. In this case we say that x is isolated in its fibre. We note that if $\psi : Z \to Y$ is a finite morphism, all fibres of ψ are finite, hence all points of X are isolated in their fibres. Now if $i : U \subset Z$ is an open immersion, we can consider the composition morphism $\psi \circ i : U \to Y$. The fibres of this morphism are also finite, hence once more all points of U are isolated in their fibres.

We shall see that given a morphism of affine schemes $\pi : X \to Y$, a point $x \in X$ is isolated in its fibre if and only if the morphism is "near x" a composition of an open immersion and a finite morphism. This is Zariski's main theorem.

Theorem 13.1 *(Zariski's main theorem)*
Let $\pi : X \to Y$ be a morphism of affine schemes and $x \in X$ such that $\dim(O_{\pi^{-1}(\pi(x)),x}) = 0$ (i.e. x is isolated in its fibre).

There exist a factorization of π through a finite morphism $\psi : Z \to Y$ and an open affine neighbourhood U of x in X such that the induced morphism $U \xrightarrow{i} Z$ is an open immersion.

Our approach to this celebrated result is completely algebraic. Indeed, it is clearly a special case of our Theorem 13.3, proved in the next section.

173

13.1   Proof of Zariski's main theorem

Definition 13.2 *Let B be an A-algebra and $\mathcal{N}$ a prime ideal of B. We say that $\mathcal{N}$ is isolated over $\mathcal{N} \cap A$ if it is maximal and minimal among prime ideals $\mathcal{P}$ of B such that $\mathcal{P} \cap A = \mathcal{N} \cap A$.*

If $\mathcal{Q} = \mathcal{N} \cap A$, put $S = A \setminus \mathcal{Q}$. Note that $\mathcal{N}$ is isolated over $\mathcal{Q}$ if and only if $\mathcal{N}S^{-1}B/\mathcal{Q}S^{-1}B$ is a maximal and minimal prime ideal of the ring $B/\mathcal{Q}S^{-1}B$.

By Theorem 8.26 (ii), we know that if $A \subset B$ is an integral extension of rings and $\mathcal{N}$ a prime ideal of B, then $\mathcal{N}$ is isolated over $\mathcal{N} \cap A$.

Theorem 13.3 *Let S be a finitely generated R-algebra. If $\mathcal{N}$ is a prime ideal of S isolated over $\mathcal{P} = \mathcal{N} \cap R$, there exist a finite R-algebra R' contained in S and an element $t \in R' \setminus (\mathcal{N} \cap R')$ such that $R'_t = S_t$.*

We first show that it is sufficient to prove the following proposition.

Proposition 13.4 *Let B be a finitely generated algebra over a subring A. Assume that A is local with maximal ideal $\mathcal{P}$ and that A is integrally closed in B. If $\mathcal{N}$ is a prime ideal of B isolated over $\mathcal{P} = \mathcal{N} \cap A$, then $A = B$.*

Assume the proposition is true. Let $S = R[x_1, ..., x_n]$ and C be the integral closure of R in S. The prime ideal $\mathcal{N}$ of S is isolated over $\mathcal{Q} = \mathcal{N} \cap C$ and $S = C[x_1, ..., x_n]$. If $T = C \setminus \mathcal{Q}$, we have, by Proposition 13.4, $C_{\mathcal{Q}} = T^{-1}B$. Hence there exist $y_i \in C$ and $t_i \in T$ such that $x_i = y_i/t_i$. If $t = \prod_1^n t_i$, it is clear that $R[y_1, ..., y_n, t]$ is a finite R-algebra such that $S_t = R[y_1, ..., y_n, t]_t$.

Proof of Proposition 13.4 in the case $B = A[x]$.
    Consider the $A/\mathcal{P}$-algebra $A[x]/\mathcal{P}A[x]$. It is naturally a quotient of $(A/\mathcal{P})[X]$. Since, by hypothesis, $\mathcal{N}/\mathcal{P}A[x]$ is a maximal and minimal prime ideal of this ring, $A[x]/\mathcal{P}A[x]$ is a strict quotient of $(A/\mathcal{P})[X]$.
    From this we deduce that there exists a relation

$$x^m + c_{m-1}x^{m-1} + ... + c_0 \in \mathcal{P}A[x],$$

with $c_i \in A$ and that every prime ideal of $A[x]$ lying over $\mathcal{P}$ is isolated.
    Denote $y = 1 + x^n + c_{n-1}x^{n-1} + ... + c_0$.
    We note that x is integral over $A[y]$, and that $\mathrm{cl}(y) = 1 \in A[x]/\mathcal{P}A[x]$. Put $\mathcal{M} = \mathcal{N} \cap A[y]$. Let us show that $\mathcal{M}$ is isolated over $\mathcal{P}$ and that $\mathrm{cl}(y) \in A[y]/\mathcal{P}A[y]$ is invertible.
    We claim that every prime ideal of $A[y]$ lying over $\mathcal{P}$ is isolated. Indeed, since $A[x]$ is integral over $A[y]$, over any chain of prime ideals of $A[y]$ there is

a chain of prime ideals of $A[x]$. Since every prime ideal of $A[x]$ lying over $\mathcal{P}$ is isolated, we are done. As a consequence, by the same argument as before, the ring $A[y]/\mathcal{P}A[y]$ is a strict quotient of $(A/\mathcal{P})[Y]$ and there is a relation

$$y^n + a_{n-1}y^{n-1} + \ldots + a_0 \in \mathcal{P}A[y],$$

with $a_i \in A$.

If $\mathrm{cl}(y) \in A[y]/\mathcal{P}A[y]$ is not invertible, it is contained in a maximal ideal of $A[y]/\mathcal{P}A[y]$. Hence there exists a maximal ideal $\mathcal{M}'$ of $A[y]$ such that $\mathcal{M}' \cap A = \mathcal{P}$ and $y \in \mathcal{M}'$. Let $\mathcal{N}'$ be a prime ideal of $A[x]$ such that $\mathcal{N}' \cap A[y] = \mathcal{M}'$. Since $\mathcal{N}' \cap A = \mathcal{P}$, we have $\mathrm{cl}(y) \in \mathcal{N}'/\mathcal{P}A[x] \subset A[x]/\mathcal{P}A[x]$, but this contradicts $\mathrm{cl}(y) = 1 \in A[x]/\mathcal{P}A[x]$.

We claim that we can assume $a_0 \notin \mathcal{P}$. Indeed, $\mathrm{cl}(y) \in A[y]/\mathcal{P}A[y]$ is not a zero divisor, hence $y(y^{n-1} + a_{n-1}y^{n-1} + \ldots + a_1) \in \mathcal{P}A[y]$ implies $y^{n-1} + a_{n-1}y^{n-1} + \ldots + a_1 \in \mathcal{P}A[y]$.

The relation can be written

$$y^n + a_{n-1}y^{n-1} + \ldots + a_0 = b_0 + b_1y + \ldots + b_sy^s,$$

with $b_i \in \mathcal{P}$. This shows that y divides $(a_0 - b_0)$. Since $(a_0 - b_0) \notin \mathcal{P}$, this element is invertible, and so is y in $A[y]$. Now if $t = \max(n, s)$, the relation

$$(a_0 - b_0)y^{-t} + (a_1 - b_1)y^{1-t} + \ldots = 0.$$

shows that $y^{-1} \in A[x]$ is integral over A. Since A is integrally closed in $A[y]$, we have $y^{-1} \in A$. But y is invertible in $A[y]$, so we have $y^{-1} \notin \mathcal{M}$, hence $y^{-1} \notin \mathcal{P}$ and y^{-1} is invertible in A. We have proved $A = A[y]$.

Since x is integral over $A[y] = A$, which is integrally closed in $A[x]$, it follows that $A = A[x]$.

Lemma 13.5 *Let $R \subset S$ be a ring extension. Assume that R is integrally closed in S, and that there is an element $t \in S$ such that S is finite over $R[t]$.*

Let $F \in R[X]$ satisfy $F(t)S \subset R[t]$, i.e. $F(t)$ is in the conductor ideal $(R[t] : S)$. If a is the leading coefficient of F, there exists an integer r such that $a^r S \subset R[t]$, i.e. $a^r \in (R[t] : S)$.

Proof Assume first that F is monic (i.e. $a = 1$). We must prove that $R[t] = S$.

If $\deg(F) = 0$, then $F = 1$ and $S = R[t]$. Assume now $\deg(F) > 0$.

If $F(t) = 0$, then t is integral over R, hence S as well, and $R = R[t] = S$.

If $F(t) \neq 0$, consider $x \in S$. There exists $G \in R[X]$ such that $xF(t) = G(t)$ and $H, Q \in R[X]$ such that $G = QF + H$, with $\deg(H) < \deg(F)$. Put $y = x - Q(t)$. We have $yF(t) - H(t) = 0$, hence $F(t) = y^{-1}H(t)$ in the ring S_y. This relation shows that t is integral over the ring $R[y^{-1}] \subset S_y$. Since $y \in S$ is integral over $R[t]$, it is integral over $R[y^{-1}]$ as well. By multiplying the relation of integral dependence of y over $R[y^{-1}]$ by a convenient power of y, we see that y is integral over R, hence in R. But $y = x - Q(t) \in R$ implies $x \in R[t]$ and $R[t] = S$, as required.

If F is not monic, we have, by the preceding case, $R_a[t] = S_a$. Let $(x_1, ..., x_n)$ be a system of generators of S as an $R[t]$-module. There is an integer r such that $a^r x_i \in R[t]$. This implies $a^r S \subset R[t]$ and the lemma is proved. $\square$

Lemma 13.6 *Let C be a subring of B such that $A \subset C \subset B$. Then $\mathcal{M} = \mathcal{N} \cap C$ is a maximal ideal of C.*

Proof Since $\mathcal{N}$ is isolated over the maximal ideal $\mathcal{P}$ of A, it is necessarily maximal so the field extension $A/\mathcal{P} \subset B/\mathcal{N}$ is finite. The double inclusion $A/\mathcal{P} \subset C/\mathcal{M} \subset B/\mathcal{N}$ shows then that $C/\mathcal{M}$ is finite over the field $A/\mathcal{P}$. Hence $C/\mathcal{M}$ is a field and $\mathcal{M}$ is a maximal ideal of C. $\square$

Proof of Proposition 13.4 when there exists $x \in B$ such that B is finite over $A[x]$.

Let $\mathcal{I} = (A[x] : B)$ be the conductor of B in $A[x]$, i.e. $\mathcal{I} = \{f \in A[x], \; fB \subset A[x]\}$. Note that $\mathcal{I}$ is an ideal of $A[x]$ and of B as well. Consider the ideal $\mathcal{M} = \mathcal{N} \cap A[x]$ of the ring $A[x]$, which is maximal by Lemma 13.6.

If $\mathcal{I} \not\subset \mathcal{M}$, we have $A[x]_\mathcal{M} = B_\mathcal{N}$ and $\mathcal{M}$ is isolated over $\mathcal{P}$. Indeed, $\mathcal{M}$ is maximal and $A[x]_\mathcal{M} = B_\mathcal{N}$ shows that $\mathcal{M}$ is minimal among prime ideals lying over $\mathcal{P}$. By the case previously treated, this shows $A = A[x]$. Since B is integral over $A[x]$, the proposition is proved in this case.

We prove now that $\mathcal{I} \subset \mathcal{M}$ is impossible.

Let $\mathcal{Q}$ be a prime ideal of B such that $\mathcal{I} \subset \mathcal{Q} \subset \mathcal{M}$ and that $\mathcal{Q}$ is minimal among prime ideals containing $\mathcal{I}$. Since B was not assumed Noetherian, the existence of such a minimal prime $\mathcal{Q}$ has not been established, but this is an easy consequence of Zorn's lemma that we can skip here. Denoting $A', B', x', \mathcal{N}', ...$ the images of $A, B, x, \mathcal{N}, ...$ in $B/\mathcal{Q}$, we show that x' is transcendental over A'.

If not there exists a non-trivial polynomial $G' \in A'[X]$ such that $G'(x') = 0$. In other words, there exists $G \in A[X]$, with dominant coefficient $a \notin \mathcal{Q}$, such that $G(x) \in \mathcal{Q}$. By Proposition 7.63, there exist $y \notin \mathcal{Q}$ and an integer r such that $yG(x)^r \in \mathcal{I}$. We have therefore $yG(x)^r B \subset A[x]$, showing that $G(x)^r$ is contained in the conductor of the ring $A[x][yB]$ in $A[x]$. Clearly $A[x][yB]$ is a finitely generated $A[x]$-module in which A is integrally closed. By Lemma

13.5 there is an integer s such that $a^s y B \subset A[x]$. This implies $a^s y \in \mathcal{I} \subset \mathcal{Q}$. Since $y \notin \mathcal{Q}$, this proves $a^s \in \mathcal{Q}$, hence a contradiction. We have proved that x' is transcendental over A'. To conclude our proof, it is sufficient to prove the following general lemma.

Lemma 13.7 *Let B be a domain, $x \in B$ and A a subring of B such that x is transcendental over A and that B is integral over $A[x]$. A prime ideal $\mathcal{N}$ of B is not isolated over $\mathcal{N} \cap A$.*

Proof We assume first that A is integrally closed. Then $A[x]$ is integrally closed by Proposition 8.19. Put $\mathcal{N}_1 = \mathcal{N} \cap A[x]$.

Since $A[x]$ is a polynomial ring over A, the prime ideal $\mathcal{N}_1$ cannot be isolated over $\mathcal{N}_1 \cap A = \mathcal{N} \cap A$. Applying the going-up or the going-down theorem, depending on whether $\mathcal{N}_1$ is minimal or maximal among prime ideals over $\mathcal{N} \cap A$, we see that $\mathcal{N}$ is not isolated over $\mathcal{N} \cap A$.

If A is not integrally closed, let A' and B' be the integral closures of A and B. Clearly x is transcendental over A' and B' is integral over $A'[x]$. If $\mathcal{N}'$ is a prime ideal of B' such that $\mathcal{N}' \cap B = \mathcal{N}$, it is not isolated over $\mathcal{N}' \cap A'$. From this fact, we easily deduce that $\mathcal{N}$ is not isolated over $\mathcal{N} \cap A = \mathcal{N}' \cap A' \cap A$. □

Proof of Proposition 13.4 in the general case

Consider $x_1, ..., x_n \in B$ such that B is finite over $A[x_1, ..., x_n]$. We proceed by induction on n (the proposition is obvious for $n = 0$).

Let C be the integral closure of $A[x_1, ..., x_{n-1}]$ in B. Since B is finite over $C[x_n]$ and $\mathcal{N}$ is isolated over $\mathcal{M} = \mathcal{N} \cap C$, we deduce from the previously treated case that $B_{\mathcal{N}} = C_{\mathcal{M}}$.

Let us show that there exists a finite $A[x_1, ..., x_{n-1}]$-algebra B' contained in C and such that

$$B'_{\mathcal{N}'} = C_{\mathcal{M}} \quad \text{for} \quad \mathcal{N}' = \mathcal{N} \cap B' = \mathcal{M} \cap B'.$$

Consider $z_1, ..., z_r \in B$ generating B as an $A[x_1, ..., x_n]$-module . Since

$$B = A[x_1, ..., x_{n-1}][x_n, z_1, ..., z_r] \quad \text{and} \quad B_{\mathcal{N}} = C_{\mathcal{M}},$$

there exists $t \in C$, with $t \notin \mathcal{M}$ such that $tx_n, tz_1, ..., tz_r \in C$. If B' is the finite $A[x_1, ..., x_{n-1}]$-algebra generated by $t, tx_n, tz_1, ..., tz_r$, putting $\mathcal{N}' = \mathcal{M} \cap B'$, we have

$$B'_{\mathcal{N}'} = C_{\mathcal{M}} = B_{\mathcal{N}}.$$

To conclude the induction on n, it is sufficient to show that $\mathcal{N}'$ is isolated over $\mathcal{P} = \mathcal{N}' \cap A$.

By Lemma 13.6, $\mathcal{N}'$ is maximal, hence maximal among prime ideals lying over $\mathcal{P}$. But $B'_{\mathcal{N}'} = B_{\mathcal{N}}$ shows clearly that it is minimal among these ideals as well.

Thus Proposition 13.4 is proved and Zariski's main theorem as well.

13.2   A factorization theorem

As an easy consequence of Theorem 13.1, we get the following more precise result.

Theorem 13.8 *Let $\pi : X \to Y$ be a morphism of affine schemes.*

(i) *The set $V \subset X$ of points x such that $\dim(O_{\pi^{-1}(\pi(x)),x}) = 0$ is open.*

(ii) *There exists an affine scheme Z and a factorization $X \xrightarrow{\phi} Z \xrightarrow{\psi} Y$ of π, satisfying the following conditions:*
    (a) the morphism ψ is finite;
    (b) the restriction morphism $\phi\mid_V: V \to Z$ is an open immersion such that $\phi^{-1}(\phi(V)) = V$.

Proof Put $A = \Gamma(Y, O_Y)$ and $B = \Gamma(X, O_X)$ and let C be the integral closure of A in B.

Consider $V \subset X$, the set of points x such that $\dim(O_{\pi^{-1}(\pi(x)),x}) = 0$. By the main theorem, if $x \in V$, there exists $t \in C$ such that $t(x) \neq 0$ and $C_t = B_t$. Obviously, the open affine subscheme $\mathrm{Spec_m}(B_t)$ of X is a neighbourhood of x contained in V. Hence V is open. Since B is Noetherian, there exists $t_1, ..., t_n \in C$ such that $V = \bigcup_i \mathrm{Spec_m}(B_{t_i})$ and that $B_{t_i} = C_{t_i}$ for all i. We recall that B is a finitely generated A-algebra, say $B = A[x_1, ..., x_r]$. Clearly, for $m \gg 0$ we have $t_i^m x_j \in C$ for all i and j. Consequently, $D = A[t_1, ..., t_n, (t_i^m x_j)_{i,j}]$ is a finite A-algebra such that $D_{t_i} = B_{t_i}$ for all i.

We put $Z = \mathrm{Spec_m}(D)$. The morphisms ψ and ϕ are induced by the homomorphisms $A \to D \to B$. We claim that the morphism $V \to Z$ is an open immersion such that $\phi^{-1}(\phi(V)) = V$. Indeed $V = \bigcup_i \mathrm{Spec_m}(B_{t_i}) \simeq \bigcup_i \mathrm{Spec_m}(D_{t_i}) \subset Z$. $\square$

13.3   Chevalley's semi-continuity theorem

Let A be a ring. Assume for the sake of simplicity that A contains an infinite field. As a consequence, for each prime ideal $\mathcal{N}$ of A, the field $k(\mathcal{N}) = A_{\mathcal{N}}/\mathcal{N}A_{\mathcal{N}}$ is infinite.

Consider a finitely generated A-algebra B, a prime ideal $\mathcal{P}$ of B, the prime ideal $\mathcal{Q} = \mathcal{P} \cap A$ of A and $S = A \setminus \mathcal{Q}$. Then $k(\mathcal{Q}) \otimes_A B = S^{-1}B/\mathcal{Q}S^{-1}B$ is a finitely generated algebra over the infinite field $k(\mathcal{Q})$. The prime ideals of this $k(\mathcal{Q})$-algebra are in natural bijection with the prime ideals of B lying over $\mathcal{Q}$. In particular $S^{-1}\mathcal{P}/\mathcal{Q}S^{-1}B$ is one of these.

Definition 13.9 *The relative dimension in $\mathcal{P}$ of B over A is*

$$\dim_{\mathcal{P}}(B, A) = \dim(B_{\mathcal{P}}/\mathcal{Q}B_{\mathcal{P}}) + \dim(S^{-1}B/\mathcal{P}S^{-1}B),$$

in other words $\dim_{\mathcal{P}}(B, A)$ is the length of a chain, of maximal length, of prime ideals of B lying over $\mathcal{Q}$ and passing through $\mathcal{P}$.

Example 13.10 *If $B = A[X_1, ..., X_r]$, then*

$$\dim_{\mathcal{P}}(B, A) = r$$

for all prime ideals $\mathcal{P}$ of B.

This is clear since $k(\mathcal{Q}) \otimes_A B$ is a polynomial ring in r variables over the field $k(\mathcal{Q})$.

Lemma 13.11 *Let A be a ring containing an infinite field, B a finitely generated A-algebra and $x_1, ..., x_r \in B$ such that B is integral over $A[x_1, ..., x_r]$. Then for all prime ideals $\mathcal{P}$ of B, we have*

$$\dim_{\mathcal{P}}(B, A) \leq r.$$

Proof Put $\mathcal{Q} = \mathcal{P} \cap A$. Let $(\mathcal{P}_i)$ be a strict chain of prime ideals of B lying over $\mathcal{Q}$. If $\mathcal{N}_i = \mathcal{P}_i \cap A[x_1, ..., x_r]$, then $(\mathcal{N}_i)$ is a strict chain of prime ideals of $A[x_1, ..., x_r]$ lying over $\mathcal{Q}$. If $\mathcal{N} = \mathcal{P} \cap A[x_1, ..., x_r]$, this shows $\dim_{\mathcal{P}}(B, A) \leq \dim_{\mathcal{N}}(A[x_1, ..., x_r], A)$.

Now consider the natural surjective homomorphism $\pi : A[X_1, ..., X_r] \to A[x_1, ..., x_r]$ and put $\mathcal{N}' = \pi^{-1}(\mathcal{N})$. Clearly,

$$\dim_{\mathcal{N}}(A[x_1, ..., x_r], A) \leq \dim_{\mathcal{N}'}(A[X_1, ..., X_r], A) = r,$$

hence we are done. $\square$

Theorem 13.12 *(The semi-continuity theorem)*
Let A be a ring containing an infinite field and B a finitely generated A-algebra. The function
$$\mathcal{P} \to \dim_{\mathcal{P}}(B, A),$$
defined in $\operatorname{Spec}(B)$, is upper semi-continuous.

Proof Consider $\mathcal{P}$ a prime ideal of B and $r = \dim_{\mathcal{P}}(B, A)$.

If $\mathcal{Q} = \mathcal{P} \cap A$, we put $S = A - \mathcal{Q}$ and $k = k(\mathcal{Q}) = A_{\mathcal{Q}}/\mathcal{Q}A_{\mathcal{Q}}$.

Next we denote by R the ring $S^{-1}B/\mathcal{Q}S^{-1}B$ and by $\mathcal{N}$ its prime ideal $\mathcal{P}S^{-1}B/\mathcal{Q}S^{-1}B$. We have

$$r = \dim(R_{\mathcal{N}}) + \mathrm{trdeg}_k(R/\mathcal{N}).$$

By 10.35, there exists $t \in B$, with $t \notin \mathcal{P}$, such that $\dim(R_t) = r$.

But R_t is a finitely generated k-algebra. By the normalization lemma, there exist $z_1, ..., z_r \in R_t$ such that R_t is integral over $k[z_1, ..., z_r]$. Since $R_t = S^{-1}B_t/\mathcal{Q}S^{-1}B_t$, there exist $x_i \in B_t$ and $s_i \in S$ such that $z_i = \mathrm{cl}(x_i/s_i) \in R_t$. Clearly R_t is integral over $k[\mathrm{cl}(x_1), ..., \mathrm{cl}(x_r)]$. In other words, we can assume $z_i = \mathrm{cl}(x_i)$.

Since R_t is integral over $k[z_1, ..., z_r]$, the prime ideal $\mathcal{N}R_t$ is isolated over $\mathcal{N}R_t \cap k[z_1, ..., z_r]$. This implies that $\mathcal{P}B_t$ is isolated over $\mathcal{P}B_t \cap A[x_1, ..., x_r]$.

By Zariski's main theorem, there exist a finite $A[x_1, ..., x_r]$-algebra $D \subset B_t$, and an element $u \in D$, with $u \notin \mathcal{P}B_t$ such that $(B_t)_u = D_u$. But $u = a/t^l$ for some $a \in B$, with $a \notin \mathcal{P}$. Put $s = at$. We have $s \notin \mathcal{P}$, hence $D(s) = \mathrm{Spec}(B) - V(sB)$ is an open neighbourhood of $\mathcal{P}$ in $\mathrm{Spec}(B)$.

We claim that for all prime ideals $\mathcal{P}' \in D(s)$, then

$$r \geq \dim_{\mathcal{P}'}(B, A).$$

But $s \notin \mathcal{P}'$ implies $t \notin \mathcal{P}'$ and $a \notin \mathcal{P}'$, hence $u \notin \mathcal{P}'B_t$. If $\mathcal{N}' = \mathcal{P}'B_t \cap D$, we have

$$B_{\mathcal{P}'} = D_{\mathcal{N}'} \quad \text{and} \quad B_{\mathcal{P}'}/\mathcal{P}'B_{\mathcal{P}'} = D_{\mathcal{N}'}/\mathcal{N}'D_{\mathcal{N}'}.$$

Put $\mathcal{Q}' = \mathcal{P}' \cap A = \mathcal{N}' \cap A$. Then,

$$\dim_{\mathcal{P}'}(B, A) = \dim(B_{\mathcal{P}'}/\mathcal{Q}'B_{\mathcal{P}'}) + \mathrm{trdeg}_{k(\mathcal{Q}')}(B_{\mathcal{P}'}/\mathcal{P}'B_{\mathcal{P}'}) =$$
$$\dim(D_{\mathcal{N}'}/\mathcal{Q}'D_{\mathcal{N}'}) + \mathrm{trdeg}_{k(\mathcal{Q}')}(D_{\mathcal{N}'}/\mathcal{N}'D_{\mathcal{N}'}) = \dim_{\mathcal{N}'}(D, A).$$

Hence, we need to prove $r \geq \dim_{\mathcal{N}'}(D, A)$, but this is Lemma 13.11, so we are done. $\qquad\square$

Corollary 13.13 *Let $\pi : X \to Y$ be a morphism of affine schemes. The function $\dim(O_{\pi^{-1}(\pi(x)),x})$, defined over X with values in $\mathbb{Z}$, is upper semicontinuous.*

Proof This is a special case of Theorem 13.12. Indeed, let $\mathcal{M}_x$ be the maximal ideal of $B = \Gamma(X, O_X)$ corresponding to x. Put $A = \Gamma(Y, O_Y)$. We have

$$O_{\pi^{-1}(\pi(x)),x} = (B/(\mathcal{M}_x \cap A)B)_{\mathcal{M}_x} \quad \text{and} \quad \mathbb{C} = B/\mathcal{M}_x = A/(\mathcal{M}_x \cap A),$$

hence

$$\dim(O_{\pi^{-1}(\pi(x)),x}) = \dim_{\mathcal{M}_x}(B, A).$$

$\qquad\square$

13.4 Exercises

1. Let A be a UFD and $B = A[x]$ a domain contained in the fraction field of A. Show that there exist $a, b \in A$ such that $B \simeq A[X]/(aX - b)$. Show that a prime ideal $\mathcal{N}$ of B is not isolated over $\mathcal{N} \cap A$ if and only if $a, b \in \mathcal{N}$.

2. Let $\pi : X \to Y$ be a rational morphism of varieties. Assume that $\dim Y = 1$ and show that all fibres of π are equidimensional of dimension equal to $\dim X - 1$.

3. Let $\pi : X \to Y$ be a rational morphism of affine varieties of dimension one. Assume that the ring $\Gamma(X, O_X)$ is integrally closed. Let B be the integral closure of $\Gamma(Y, O_Y)$ in the fraction field of $\Gamma(X, O_X)$. Show that there exist elements s and t in B such that $\Gamma(X, O_X) = B_s \cap B_t$.

4. Consider the ring extensions $\mathbb{C}[X, Y] \subset \mathbb{C}[X, Y, XZ, YZ] \subset \mathbb{C}[X, Y, Z]$ and the corresponding morphisms of affine schemes

$$\mathbb{A}_3 \to \operatorname{Spec_m}(\mathbb{C}[X, Y, XZ, YZ]) \to \mathbb{A}_2.$$

Study the dimension of the fibres of these morphisms.

14

Integrally closed Noetherian rings

We begin to prepare the reader for the theory of divisors. Integrally closed Noetherian rings, often called normal rings, carry Weil divisors. In the first part of this chapter our main resut is "Serre's criterion", Theorem 14.4. We decided to break it into pieces: we characterize separately reduced Noetherian rings, Theorem 14.1, and Noetherian domains, Theorem 14.2. In the second and third sections we study discrete valuations and their rings: along with Dedekind rings they are at the centre of the theory of divisors.

In each of the two last sections we present a useful theorem. In the fourth, we prove that under certain hypotheses, always satisfied in complex geometry, the integral closure of a Noetherian domain is finite over the domain. In the fifth section, we study the integral closure of a domain in a finite normal extension of its field of fractions: we focus on the action of the Galois group on the set of prime ideals of this integral closure.

14.1   Reduced Noetherian rings

Theorem 14.1 *If A is a Noetherian ring the following conditions are equivalent:*

(i) *the ring A is reduced, i.e. $\mathrm{Nil}(A) = (0)$;*

(ii) *for all prime ideals $\mathcal{P} \in \mathrm{Ass}(A)$, the local ring $A_{\mathcal{P}}$ is a field;*

(iii) *for all prime ideals $\mathcal{P} \in \mathrm{Ass}(A)$, the local ring $A_{\mathcal{P}}$ is a domain.*

Proof If A is reduced, (0) is the intersection of the minimal prime ideals of A. There are finitely many minimal prime ideals by Theorem 3.30, say $\mathcal{P}_i$ with $i = 1, ..., n$. Then $(0) = \bigcap_1^n \mathcal{P}_i$ is a minimal primary decomposition. This shows that $\mathrm{Ass}(A) = \{\mathcal{P}_1, ..., \mathcal{P}_n\}$. By Corollary 7.59 (i), we have

$$0A_{\mathcal{P}_i} = \mathcal{P}_i A_{\mathcal{P}_i}$$

for $i = 1, ..., n$. Hence $A_{\mathcal{P}}$ is a field for all $\mathcal{P} \in \mathrm{Ass}(A)$ and (i) implies (ii). We note that (ii) implies (iii) obviously.

To show that (iii) implies (i), consider a nilpotent element $a \in A$. If $\mathcal{P} \in \mathrm{Ass}(A)$, then $a/1 \in A_{\mathcal{P}}$ is nilpotent. Since $A_{\mathcal{P}}$ is a domain, this shows that $aA_{\mathcal{P}} = (0)$ for all $\mathcal{P} \in \mathrm{Ass}(A)$. Consequently, $\mathrm{Ass}(A) \cap \mathrm{Ass}(aA) = \emptyset$. Since $\mathrm{Ass}(aA) \subset \mathrm{Ass}(A)$, this proves $\mathrm{Ass}(aA) = \emptyset$, hence $aA = (0)$.   □

Theorem 14.2 *Let A be a Noetherian ring which is not the product of two rings. The following conditions are equivalent:*

(i) *the ring A is a domain;*

(ii) *for all $x \in A$ and all $\mathcal{N} \in \mathrm{Ass}(A/xA)$, the local ring $A_{\mathcal{N}}$ is a domain.*

Proof A fraction ring of a domain is a domain, hence (i) implies (ii).

Conversely, A is reduced by Theorem 14.1. We assume that A is not a domain and find $x \in A$ and $\mathcal{N} \in \mathrm{Ass}(A/xA)$ such that $A_{\mathcal{N}}$ is not a domain.

Consider the minimal primary decomposition $(0) = \bigcap_1^n \mathcal{P}_i$, with $n > 1$, of (0) in A. If we put $\mathcal{P} = \mathcal{P}_1$ and $\mathcal{Q} = \bigcap_2^n \mathcal{P}_i$, we have $(0) = \mathcal{P} \cap \mathcal{Q}$.

If $(\mathcal{P} + \mathcal{Q}) = A$, then $\mathcal{P}$ and $\mathcal{Q}$ are comaximal and $A \simeq A/\mathcal{P} \times A/\mathcal{Q}$ by Theorem 1.62. This contradicts our hypothesis, hence $(\mathcal{P} + \mathcal{Q})$ is an ideal of A.

If $\mathcal{N} \in \mathrm{Ass}(A/(\mathcal{P} + \mathcal{Q}))$, it is clear that $A_{\mathcal{N}}$ is not a domain. Indeed, since $\mathcal{Q} \subset \mathcal{N}$ there exists $i \geq 2$ such that $\mathcal{P}_i \subset \mathcal{N}$, hence $\mathcal{N}$ contains the minimal prime ideals $\mathcal{P}_1$ and $\mathcal{P}_i$. Moreover we claim that there exists $x \in A$ such that $\mathcal{N} \in \mathrm{Ass}(A/xA)$.

There exist $a \in \mathcal{P}$ such that $a \notin \mathcal{P}_i$ for $i > 1$ and $b \in \mathcal{Q}$ such that $b \notin \mathcal{P}$. We obviously have

$$aA \cap bA \subset \mathcal{P} \cap \mathcal{Q} = (0), \quad (0) : a = \bigcap_2^n \mathcal{P}_i = \mathcal{Q} \quad \text{and} \quad (0) : b = \mathcal{P}.$$

We claim that $\mathcal{N} \in \mathrm{Ass}(A/(a + b)A)$. To begin with, we show that $(a + b)A : a = (\mathcal{P} + \mathcal{Q})$:

$$ax = (a + b)y \iff a(x - y) = by \iff a(x - y) = by = 0.$$

But this is equivalent to

$$(x - y) \in \mathcal{Q} \quad \text{and } y \in \mathcal{P} \quad \text{hence to} \quad x = (x - y) + y \in (\mathcal{P} + \mathcal{Q}).$$

Now since $\mathcal{N} \in \mathrm{Ass}(A/(\mathcal{P} + \mathcal{Q}))$, there exists $c \in A$ such that $\mathcal{N} = (\mathcal{P} + \mathcal{Q}) : c$. Then

$$acz \in (a + b)A \iff cz \in (\mathcal{P} + \mathcal{Q}) \iff z \in \mathcal{N},$$

hence $\mathcal{N} = (a + b)A : ac$, and $\mathcal{N} \in \mathrm{Ass}(A/(a + b)A)$. The theorem is proved.   □

14.2 Integrally closed Noetherian rings

We recall that a prime ideal $\mathcal{P}$ of a Noetherian ring A is of height one if all prime ideals of $A_{\mathcal{P}}$ except $\mathcal{P}A_{\mathcal{P}}$ are minimal.

Lemma 14.3 *Let R be a local Noetherian domain which is not a field and $\mathcal{M}$ its maximal ideal. The following conditions are equivalent:*

(i) *the ring R is integrally closed and there exists $x \in R$ such that $\mathcal{M} \in$ Ass(R/xR);*

(ii) *the maximal ideal $\mathcal{M}$ is principal;*

(iii) *the ring R is a principal ideal ring;*

(iv) *the ring R is integrally closed of dimension 1.*

Proof Assume (i). There exists $a \in R$ such that $\mathcal{M} = (xR : a)$.

If $a\mathcal{M} \subset x\mathcal{M}$, we have $(a/x)\mathcal{M} \subset \mathcal{M}$. This shows that $\mathcal{M}$ is a faithful $R[a/x]$-module which is finitely generated as an R-module. Consequently, a/x is integral over R by Theorem 8.5. Since R is integrally closed, $a/x \in R$, hence $a \in xR$. This contradicts $\mathcal{M} = (xR : a)$, hence we know $a\mathcal{M} \not\subset x\mathcal{M}$. This implies $x \in a\mathcal{M}$, so there exists $b \in \mathcal{M}$ such that $x = ab$. Then $\mathcal{M} = (xR : a) = (abR : aR) = bR$ and (ii) is proved.

Assume now $\mathcal{M} = xR$. Let us show that R is a principal ideal ring. If not, let $\mathcal{I}$ be an ideal of R maximal among non-principal ideals. Since $\mathcal{I} \subset \mathcal{M} = xR$, there is an ideal $\mathcal{J}$ such that $\mathcal{I} = x\mathcal{J}$. We claim that $\mathcal{I} \neq \mathcal{J}$. Indeed $\mathcal{I} = x\mathcal{I}$ implies $\mathcal{I} \subset \mathcal{M}\mathcal{I}$ and $\mathcal{I} = (0)$ by Nakayama's lemma. Hence $\mathcal{J}$ is principal. But if $\mathcal{J} = aR$, then $\mathcal{I} = xaR$, a contradiction. Hence (ii) $\Longrightarrow$ (iii).

A principal ideal ring is integrally closed and its non-zero prime ideals are maximal, hence (iii) $\Rightarrow$ (iv).

To conclude, assume (iv). If $x \in \mathcal{M}$ is non-zero, then $\mathcal{M}$ is a minimal prime ideal of xR. This proves $\mathcal{M} \in$ Ass(R/xR) and (i). $\square$

Theorem 14.4 *(Serre's criterion)*
Let A be a Noetherian ring which is not a product of two rings. The following conditions are equivalent:

(i) *the ring A is integrally closed;*

(ii) *for all $x \in A$ and all $\mathcal{N} \in$ Ass(A/xA), the local ring $A_{\mathcal{N}}$ is a field or a principal ideal ring.*

(iii) *for all $x \in A$ and all $\mathcal{N} \in$ Ass(A/xA), the local ring $A_{\mathcal{N}}$ is integrally closed.*

Proof Assume (i). Consider a non-zero element $x \in A$ and $\mathcal{N} \in \mathrm{Ass}(A/xA)$. Then $A_{\mathcal{N}}$ is an integrally closed local ring such that its maximal ideal $\mathcal{N}A_{\mathcal{N}}$ is associated to $A_{\mathcal{N}}/xA_{\mathcal{N}}$. By Lemma 14.3, $A_{\mathcal{N}}$ is a principal ideal ring and (ii) is proved.

Since a principal ideal ring is integrally closed, (ii) implies (iii).

Assume (iii). By Theorem 14.2, A is a domain. If a domain is an intersection of integrally closed rings it is obviously integrally closed. Hence the following general lemma proves (i).

Lemma 14.5 *Let A be a domain. If $E \subset \mathrm{Spec}(A)$ is the set of prime ideals $\mathcal{N}$ such that there exists $x \in A$ with $\mathcal{N} \in \mathrm{Ass}(A/xA)$, then*

$$A = \bigcap_{\mathcal{N} \in E} A_{\mathcal{N}}.$$

Proof It is clear that $A \subset \bigcap_{\mathcal{N} \in E} A_{\mathcal{N}}$.

If $y/x \in \bigcap_{\mathcal{N} \in E} A_{\mathcal{N}}$, we have $y/x \in \bigcap_{\mathcal{N} \in \mathrm{Ass}(A/xA)} A_{\mathcal{N}}$. Hence $y \in xA_{\mathcal{N}}$ for all $\mathcal{N} \in \mathrm{Ass}(A/xA)$. In other words $\mathrm{cl}(y) \in A/xA$ satisfies $\mathrm{cl}(y)(A/xA)_{\mathcal{N}} = (0)$ for all $\mathcal{N} \in \mathrm{Ass}(A/xA)$. This proves $(0 : \mathrm{cl}(y)) \not\subset \mathcal{N}$ for all $\mathcal{N} \in \mathrm{Ass}(A/xA)$, hence $\mathrm{cl}(y) = 0$. This means $y \in xA$, hence $y/x \in A$ and the lemma follows. □

Our two last results in this section are almost immediate now, but we choose to present them separately for later reference.

Proposition 14.6 *If R is integrally closed, then*

$$R = \bigcap_{\mathrm{ht}(\mathcal{P})=1} R_{\mathcal{P}}.$$

Proof By Lemma 14.5, this is a consequence of the following lemma.

Lemma 14.7 : *Let R be an integrally closed Noetherian domain and $\mathcal{P}$ a prime ideal of R. The following conditions are equivalent:*

(i) *the height of $\mathcal{P}$ is one;*

(ii) *there exists a non-zero element $x \in R$ such that $\mathcal{P} \in \mathrm{Ass}(R/xR)$;*

(iii) *the local ring $R_{\mathcal{P}}$ is a principal ideal ring.*

Proof If $\mathrm{ht}(\mathcal{P}) = 1$ and $x \in \mathcal{P}$ is non-zero, then $\mathcal{P}$ is a minimal prime of xR. This proves $\mathcal{P} \in \mathrm{Ass}(R/xR)$, hence (i) implies (ii).

If $\mathcal{P} \in \mathrm{Ass}(R/xR)$, then $R_{\mathcal{P}}$ is a principal ideal ring, by Theorem 14.4, and (ii) implies (iii).

In a principal ideal ring all non-zero prime ideals have height 1, hence $\mathrm{ht}(\mathcal{P}) = 1$. □

14.3 Discrete valuation rings. Dedekind rings

Definition 14.8 *A local principal ideal ring is called a discrete valuation ring (DVR).*

Proposition 14.9 *A Noetherian local domain R, with maximal ideal $\mathcal{M}$, is a discrete valuation ring if and only if the $R/\mathcal{M}$-vector space $\mathcal{M}/\mathcal{M}^2$ has rank one.*

Proof If $a \in \mathcal{M}$ is such that $\mathrm{cl}(a) \in \mathcal{M}/\mathcal{M}^2$ generates the $R/\mathcal{M}$-vector space $\mathcal{M}/\mathcal{M}^2$, then $\mathcal{M} = aR$ by Nakayama's lemma. The converse is obvious. □

Exercise 14.10 Let $P \in \mathbb{C}[X, Y] = R$ be non-zero and $(a, b) \in \mathbb{C}^2$ be such that $P(a, b) = 0$. Put $\mathcal{M} = (X - a, Y - b)$ and show that the following conditions are equivalent:

(i) $(R/PR)_{\mathcal{M}}$ is a discrete valuation ring;

(ii) $P \notin \mathcal{M}^2$;

(iii) $(\partial P/\partial X, \partial P/\partial Y) \not\subset \mathcal{M}$;

(iv) there exist $(c, d) \in \mathbb{C}^2$ such that $c(\partial P/\partial X)(a, b) + d(\partial P/\partial Y)(a, b) \neq 0$.

Definition 14.11 *If uV is the maximal ideal of the discrete valuation ring V, we say that u is a uniformizing element for V.*

Proposition 14.12 *Let V be a discrete valuation ring and u a uniformizing element for V. For any non-zero element x in the fraction field $K(V)$ of V, there exists a unique integer n such that $xV = u^n V$.*

Proof Clearly, uV is the unique non-trivial prime ideal of the principal ideal ring V. Therefore, $f \in V$ is irreducible if and only if $f = us$, with s invertible. This shows that a non-zero element $a \in V$ has a unique decomposition $a = u^n s$, with s invertible. Consider $x = a/b \in K(V)$, with $a, b \in V$. If $a = u^n s$ and $b = u^m s'$, with s and s' invertible elements of V, it is clear that $xV = u^{n-m}V$. Finally, if $u^k V = u^l V$, with $k \geq l$, we find $u^{k-l}V = V$, hence $k = l$. □

Note that the map $v : K(V)^* \to \mathbb{Z}$ which associates to $x \in K(V)^*$ the unique integer n such that $xV = u^n V$ satisfies the following conditions:

$$(i) \quad v(xy) = v(x) + v(y), \quad (ii) \quad v(x + y) \geq \min(v(x), v(y))$$
$$(iii) \quad v(x) \geq 0 \iff x \in V.$$

Definition 14.13 *A discrete valuation on a field K is a surjective homomorphism $v : K^* \to \mathbb{Z}$, from the multiplicative group K^* to the additive group $\mathbb{Z}$, such that $v(a + b) \geq \min(v(a), v(b))$.*

Exercise 14.14 Let $p \in \mathbb{Z}$ be a prime number. For $a \in \mathbb{Z} - \{0\}$, put $v_p(a) = \max(n, \ a \in p^n\mathbb{Z})$ and for $q = a/b \in \mathbb{Q}^*$, where a and b are relatively prime, put $v_p(q) = v_p(a) - v_p(b)$. Show that v_p is a discrete valuation on $\mathbb{Q}$.

Theorem 14.15

(i) *If V is a discrete valuation ring, there exists a discrete valuation v on the fraction field K of V such that $V - \{0\}$ is the set of all $a \in K^*$ such that $v(a) \geq 0$.*

(ii) *Let v be a discrete valuation on a field K. The set R formed by $\{0\}$ and the elements $a \in K^*$ such that $v(a) \geq 0$ is a discrete valuation ring.*

Proof (i) has already been proved.

For (ii), we note first that the relation $v(a + b) \geq \min(v(a), v(b))$ shows that R is a commutative additive subgroup of K. Then from the relation $v(ab) = v(a) + v(b)$ it follows that R is a ring.

We note next that for $a, b \in R - \{0\}$, then

$$b \in aR \Leftrightarrow b/a \in R \Leftrightarrow v(b) \geq v(a).$$

This shows that if $\mathcal{I} \subset R$ is an ideal and $a \in \mathcal{I}$ is such that $v(b) \geq v(a)$ for all $b \in \mathcal{I}$, then $\mathcal{I} = aR$. Hence R is principal.

Finally, let $u \in R$ be such that $v(u) = 1$. If $y \in R$ is non-zero and non-invertible, then $y^{-1} \notin R$, hence $v(y) = -v(y^{-1}) > 0$. This shows $y \in uR$. Consequently uR is the only maximal ideal of R. $\square$

Exercise 14.16 Let R be a UFD and $f \in R$ an irreducible element. For $a \in R - \{0\}$, put $v_f(a) = \max(n, \ a) \in f^n R$. Show that we can define $v_f(a/b) = v_f(a) - v_f(b)$ for $a/b \in K(R)^*$, that v_f is a discrete valuation on the field $K(R)$ and that the discrete valuation ring associated to this valuation is $R_{(fR)}$ (we recall that fR is a prime ideal of R).

Definition 14.17 *A Noetherian domain A such that for each prime ideal $\mathcal{P}$ of A the local ring $A_{\mathcal{P}}$ is a discrete valuation ring, is called a Dedekind ring.*

Note that by definition a fraction ring of a Dedekind ring is a Dedekind ring.

Example 14.18 A principal ideal ring is a Dedekind ring.

Exercise 14.19 Let $P \in \mathbb{C}[X, Y]$ be non-zero. Use Exercise 14.10 to show that

$$\mathbb{C}[X, Y]/P\mathbb{C}[X, Y]$$

is a Dedekind ring if and only if $(P, \partial P/\partial X, \partial P/\partial Y) = \mathbb{C}[X, Y]$.

Proposition 14.20 *A Noetherian domain R is a Dedekind ring if and only if it is integrally closed and of dimension one.*

Proof Assume R is a Dedekind ring. Let $\mathcal{M}$ be a maximal ideal of R. The local ring $R_\mathcal{M}$ is principal, hence integrally closed. Since $R = \bigcap_\mathcal{M} R_\mathcal{M}$, for $\mathcal{M}$ maximal, this shows that R is integrally closed. Furthermore, if $\mathcal{M}$ is a maximal ideal of R, then $R_\mathcal{M}$ is principal, hence $\dim(R_\mathcal{M}) = 1$. This proves $\dim(R) = 1$.

Conversely, let R be an integrally closed ring such that all non trivial prime ideals are maximal. If $\mathcal{P}$ is a maximal ideal of R, then $\mathcal{P}R_\mathcal{P}$ is the unique non-trivial prime ideal of the integrally closed local ring $R_\mathcal{P}$. This ring is therefore principal, hence a discrete valuation ring, by Lemma 14.3. □

Theorem 14.21 *Let D be a Dedekind ring. Then:*

(i) *any non-trivial prime ideal of D is maximal;*

(ii) *any non-trivial primary ideal of D is a power of a maximal ideal;*

(iii) *any non-trivial ideal of D has a unique decomposition as a product of maximal ideals.*

Proof (i) is already proved in Proposition 14.20.

(ii) Let $\mathcal{Q}$ be a $\mathcal{P}$-primary ideal. Then $D_\mathcal{P}$ is a discrete valuation ring. Hence, $\mathcal{Q}D_\mathcal{P} = \mathcal{P}^n D_\mathcal{P}$ for some positive integer n by Proposition 14.12. But $\mathcal{P}^n$ is, like $\mathcal{Q}$, a $\mathcal{P}$-primary ideal, since $\mathcal{P}$ is the unique prime ideal containing $\mathcal{P}^n$. By Theorem 7.58 (i), we get

$$\mathcal{Q} = \mathcal{Q}A_\mathcal{P} \cap A = \mathcal{P}^n A_\mathcal{P} \cap A = \mathcal{P}^n.$$

(iii) Let $\mathcal{I} = \bigcap_1^r \mathcal{P}_i^{n_i}$ be a minimal primary decomposition of the non-trivial ideal $\mathcal{I}$. The prime ideals $\mathcal{P}_i$ are all maximal, hence all minimal prime ideals of $\mathcal{I}$ and the primary decomposition is uniquely defined. Since the ideals $\mathcal{P}_i^{n_i}$ are pairwise comaximal, we have

$$\mathcal{I} = \mathcal{P}_1^{n_1} ... \mathcal{P}_r^{n_r}.$$

□

Proposition 14.22 *In a Dedekind ring D, any ideal is generated by at most two elements.*

Proof Let $\mathcal{I} = \mathcal{P}_1^{n_1}...\mathcal{P}_r^{n_r}$. By the avoiding lemma, for all i, there exist $a_i \in \mathcal{P}_i^{n_i}$ such that

$$a_i \notin \mathcal{P}_j, \text{ for } j \neq i \quad \text{and} \quad a_i \notin \mathcal{P}_i^{n_i+1}.$$

Put $a = a_1 a_2 ... a_r$. We have

$$aD_{\mathcal{P}_i} = \mathcal{P}_i^{n_i} D_{\mathcal{P}_i} = \mathcal{I} D_{\mathcal{P}_i}.$$

This proves

$$aD = \mathcal{P}_1^{n_1}...\mathcal{P}_r^{n_r} \mathcal{J} = \mathcal{P}_1^{n_1} \cap ... \cap \mathcal{P}_r^{n_r} \cap \mathcal{J}, \quad \text{where} \quad \mathcal{J} \not\subset \mathcal{P}_j \text{ for } j = 1, ..., r.$$

Let $(\mathcal{P}'_l)_{1 \leq l \leq k}$ be the prime ideals associated to $\mathcal{J}$. Since $\mathcal{I} \not\subset \mathcal{P}'_l$ for all l, there exists $b \in \mathcal{I}$ such that $b \notin \mathcal{P}'_l$ for all l. Consequently,

$$(a, b)D_{\mathcal{P}_i} = \mathcal{I} D_{\mathcal{P}_i}, \text{ for } i = 1, ..., r$$

and

$$(a, b)D_{\mathcal{P}} = D_{\mathcal{P}} \quad \text{for} \quad \mathcal{P} \neq \mathcal{P}_i, \text{ for all } i,$$

which shows $(a, b)D = \mathcal{I}$. $\square$

Proposition 14.23 *If a Dedekind ring D has only finitely many maximal ideals, it is a principal ideal ring.*

Proof By Theorem 14.21, it suffices to show that any maximal ideal of D is principal. Let $(\mathcal{P}_i)_{1 \leq i \leq n}$ be the maximal ideals of D. By the avoiding lemma, there exists

$$a_i \in \mathcal{P}_i, \quad a_i \notin \mathcal{P}_j \text{ for } j \neq i \quad \text{and} \quad a_i \notin \mathcal{P}_i^2.$$

Since $a_i \notin \mathcal{P}_j$ for $j \neq i$, the principal ideal $a_i D$ is $\mathcal{P}_i$-primary. Consequently, $a_i D$ is a power of $\mathcal{P}_i$. Then, $a_i \notin \mathcal{P}_i^2$ shows $a_i D = \mathcal{P}_i$. $\square$

14.4 Integral extensions of Noetherian domains

The integral closure of a Noetherian domain A is not necessarily Noetherian. We won't let this unpleasant piece of news trouble us. When our Noetherian ring A is a finitely generated $\mathbb{C}$-algebra its integral closure is not only Noetherian but it is also finite over A. This is essentially the theme of this section.

Theorem 14.24 *Let $A \subset B$ be an integral extension of domains of characteristic zero such that the fraction field L of B is a finite extension of the fraction field K of A.*

If A is Noetherian and integrally closed, then B is finite over A.

Proof Consider $z \in L$ such that $L = K[z]$ and P the minimal polynomial of z over K. Let Ω be an algebraic closure of L, and $L' \in \Omega$ the decomposition field of P. Note that it suffices to show that the integral closure B' of A in L' is finite over A. Indeed, B is an A-submodule of B' and A is Noetherian. In other words, we can assume that the extension $K \subset L$ is normal.

Put $S = A - \{0\}$. The domain $S^{-1}B$ is integral over the field $S^{-1}A = K$, hence it is a field, in other words $S^{-1}B = L$. There exists therefore a basis $(x_1, ..., x_n)$ of the K-vector space L such that $x_i \in B$ for $i = 1, ..., n$.

Now we recall that $(x, y) \rightarrow \text{Tr}_{L/K}(xy)$ is a non-degenerate symmetric bilinear form on the K-vector space L. Next we consider the basis $(y_1, ..., y_n)$ of L dual to the basis $(x_1, ..., x_n)$ for this form. In other words, we have $\text{Tr}_{L/K}(x_i y_j) = \delta_{ij}$. We claim that B is a submodule of the A-module generated by $y_1, ..., y_n$; since A is Noetherian this proves that B is a finitely generated A-module. Consider $v \in B \subset L$. There exists $a_i \in K$ such that $v = a_1 y_1 + ... + a_n y_n$. We have

$$\text{Tr}_{L/K}(x_i v) = \sum_j a_j \text{Tr}_{L/K}(x_i y_j) = a_i.$$

But $x_i \in B$ and $v \in B$ imply $x_i v \in B$. By Lemma 8.31, this shows that the minimal polynomial of $x_i v$ over K is in $A[X]$. Consequently, $a_i = \text{Tr}_{L/K}(x_i v) \in A$ by Corollary 9.38 and we are done. $\square$

Corollary 14.25 *Let K be a field of characteristic zero, A a domain finitely generated as a K-algebra, S a multiplicatively closed part of A and $A' = S^{-1}R$.*

If L is a field finite over the fraction field of A, the integral closure B' of A' in L is finite over A'. In particular, the integral closure of A' is finite over A'.

Proof We note first that if B is the integral closure of A in L, then $S^{-1}B$ is the integral closure of A' in L. Hence it suffices to show that B is finite over R.

We know, by the normalization lemma, that A is finite over a polynomial ring R. The fraction field of A is obviously finite over the fraction field $K(R)$ of R. Hence L is finite over $K(R)$. The ring B is obviously the integral closure of R in L. Since R is integrally closed, the ring B is finite over R by Theorem 14.24. This implies that B is finite over A. $\square$

14.5   Galois group and prime ideals

In the final section of this chapter, we consider an integrally closed Noetherian
domain A, of characteristic zero, and its integral closure B in a finite normal
extension of the fraction field of A. If $\mathcal{P}$ is a prime ideal of B and $\mathcal{Q} =
\mathcal{P} \cap A$, we show that the fraction field of $B/\mathcal{P}$ is a finite normal extension of
the fraction field of $A/\mathcal{Q}$ and we relate the Galois group of this extension to
$\mathrm{Gal}(K(B)/K(A))$.

Theorem 14.26 *Let $K \subset L$ be a finite normal extension of fields of characteristic zero and $G = \mathrm{Gal}(L/K)$ its Galois group. Let $B \subset L$ be a ring such that
$g(B) \subset B$ for all $g \in G$ and $A = B \cap K = \{x \in B, \ g(x) = x \quad \text{for all } g \in G\}$.
Let $\mathcal{P}$ be a prime ideal of B such that the fraction field of $B/\mathcal{P}$ has characteristic zero and $\mathcal{Q} = \mathcal{P} \cap A$. Then:*

(i) *the ring B is integral over A;*

(ii) *if $g \in G$, then $g(\mathcal{P})$ is a prime ideal of B such that $g(\mathcal{P}) \cap A = \mathcal{Q}$ and for
each prime ideal $\mathcal{N}$ of B such that $\mathcal{N} \cap A = \mathcal{Q}$ there exists $g \in G$ such
that $\mathcal{N} = g(\mathcal{P})$;*

(iii) *the fraction field k' of $B/\mathcal{P}$ is a finite normal extension of the fraction
field k of $A/\mathcal{Q}$;*

(iv) *if G' denotes the subgroup of G formed by all $g \in G$ such that $g(\mathcal{P}) = \mathcal{P}$
(the stabilizer of $\mathcal{P}$), then there is a natural surjective homomorphism
from G' to the group $\mathrm{Gal}(k'/k)$.*

Proof   (i) Put $G = \{g_1, ..., g_n\}$. If $x \in B$, consider the polynomial

$$\prod_{i=1}^{n}(X - g_i(x)) = X^n + a_1 X^{n-1} + ... + a_n.$$

By an argument that we have now used several times, $g(a_i) = a_i$ for all $g \in G$.
This proves $a_i \in A$ for all i, and consequently that x is integral over A.

(ii) Since g is a ring automorphism, $g(\mathcal{P})$ is a prime ideal of $B = g(B)$.
Furthermore,

$$y \in g(\mathcal{P}) \cap A \Longleftrightarrow y = g^{-1}(y) \in \mathcal{P} \cap A = \mathcal{Q}.$$

Now let $\mathcal{N}$ be a prime ideal of B such that $\mathcal{N} \cap A = \mathcal{Q}$ and $x \in \mathcal{N}$. Clearly,
$g(\prod_1^n g_i(x)) = \prod_1^n g_i(x)$ for all $g \in G$. This shows $\prod_1^n g_i(x) \in \mathcal{N} \cap A = \mathcal{P} \cap A$.
Hence there exists i such that $g_i(x) \in \mathcal{P}$. This proves $x \in g_i^{-1}(\mathcal{P})$ and

$$\mathcal{N} \subset \bigcup_1^n g_i(\mathcal{P}).$$

By the avoiding lemma, there exists i such that $\mathcal{N} \subset g_i(\mathcal{P})$. Since $\mathcal{N} \cap A = \mathcal{P} \cap A = g_i(\mathcal{P}) \cap A$, we have $\mathcal{N} = g_i(\mathcal{P})$ by Theorem 8.26.

(iii) Consider $x \in B$ and denote its class in $B/\mathcal{P} \subset k'$ by $\bar{x}$. Clearly, x is a root of the polynomial $P(X) = \prod_1^n (X - g_i(x)) = X^n + a_1 X^{n-1} + \ldots + a_n \in A[X]$. Consequently, $\bar{x}$ is a root of the polynomial $\overline{P}(X) = \prod_1^n (X - \overline{g_i(x)}) \in (B/\mathcal{P})[X]$ whose coefficients are in fact in $A/(\mathcal{P} \cap A) = A/\mathcal{Q}$. This shows that the conjugates to $\bar{x}$ in an algebraic closure of k' are among the elements $\overline{g_i(x)}$, hence in k'. Now, an element $z \in k'$ is of the form $\bar{y}/\bar{s}$, with $y \in B$ and $s \in A - \mathcal{Q}$. This shows that its conjugates are among the elements $\overline{g_i(y)}/\bar{s}$, hence in k'. Consequently, the extension $k \subset k'$ is normal. Furthermore, an element of k' has at most n conjugates over k. Since k' has characteristic zero, this proves $[k' : k] \leq n$ by Theorem 9.20.

(iv) If $g(\mathcal{P}) = \mathcal{P}$, it is clear that g induces an injective $A/\mathcal{Q}$-endomorphism of $B/\mathcal{P}$, hence a k-isomorphism of k'. We have therefore defined a natural map $\pi : G' \to \mathrm{Gal}(k'/k)$ which is obviously a group homomorphism. We claim that π is surjective.

Let $(\mathcal{P}_1, ..., \mathcal{P}_r)$ be the pairwise distinct prime ideals of B such that $\mathcal{P}_i \cap A = \mathcal{P} \cap A$, with $\mathcal{P} = \mathcal{P}_1$.

Consider $x \in \bigcap_2^r \mathcal{P}_i$. If $g \notin G'$, then $g^{-1} \notin G'$ and there exists $i > 1$ such that $g(\mathcal{P}_i) = \mathcal{P}$. This implies $g(x) \in \mathcal{P}$, in other words $\overline{g(x)} = 0$. Assume $x \notin \mathcal{P}$, in other words $\bar{x} \neq 0$. The conjugates to $\bar{x}$ in k', over k, are non-zero and among the elements $\overline{g_i(x)}$. This shows that these conjugates are among the elements $\overline{g(x)}$, with $g \in G'$, in other words that G' operates transitively on the set of all conjugates of $\bar{x}$ in k'. Next we claim that there exists $x \in \bigcap_2^r \mathcal{P}_i$ such that $k' = k[\bar{x}]$ and we note that this will conclude the proof of our theorem.

Put $S = A - \mathcal{Q}$ and note that $k' = S^{-1}B/\mathcal{P}S^{-1}B$. By Theorem 9.20, there exists $z \in S^{-1}B$ such that its class $\bar{z} \in S^{-1}B/\mathcal{P}S^{-1}B = k'$ satisfies $k' = k[\bar{z}]$. Since in the ring $S^{-1}B$ the ideals $\mathcal{P}S^{-1}B$ and $(\bigcap_2^r \mathcal{P}_i)S^{-1}B$ are comaximal, there exist $x' \in (\bigcap_2^r \mathcal{P}_i)S^{-1}B$ and $y' \in \mathcal{P}S^{-1}B$ such that $z = x' + y'$. This shows $\bar{z} = \overline{x'} \in S^{-1}B/\mathcal{P}S^{-1}B = k'$. Now if $x \in \bigcap_2^r \mathcal{P}_i$ is such that $x' = x/s$, with $s \in S$, we obviously have $k' = k[\bar{x}]$. $\square$

We can now come back briefly to the Going-down theorem. Let $A \subset B$ be an integral extension of domains such that A is integrally closed. We assume here, for the sake of simplicity, that the fraction field L of B has characteristic zero and is finite over the fraction field K of A.

Let $\mathcal{Q}_1 \subset \mathcal{Q}_0$ be a chain of prime ideals of A and $\mathcal{P}_0$ a prime ideal of B such that $\mathcal{P}_0 \cap A = \mathcal{Q}_0$. We want to explain once again why there exists a prime ideal $\mathcal{P}_1 \subset \mathcal{P}_0$ in B such that $\mathcal{P}_1 \cap A = \mathcal{Q}_1$.

Consider $z \in L$ such that $L = K[z]$ and P the minimal polynomial of z over K. Let Ω be an algebraic closure of L, and $L' \in \Omega$ the decomposition field of P. Note that L' is a finite normal extension of K and consider B' the integral closure of B in L'. Now, there exists a prime ideal $\mathcal{P}'_0$ of B' such that $\mathcal{P}'_0 \cap B = \mathcal{P}_0$. We claim that it suffices to find a prime ideal $\mathcal{P}'_1 \subset \mathcal{P}'_0$ in B', such that $\mathcal{P}'_1 \cap A = \mathcal{Q}_1$. Indeed, the prime ideal $\mathcal{P}_1 = \mathcal{P}'_1 \cap B$ satisfies $\mathcal{P}_1 \subset \mathcal{P}_0$ and $\mathcal{P}_1 \cap A = \mathcal{Q}_1$.

First we note that for each $g \in G = \mathrm{Gal}(L'/K)$, we have $g(B') = B'$ and that $B' \cap K = A$. We assume furthermore that the fraction field of $B'/\mathcal{P}'_0$ has characteristic zero and we apply Theorem 14.26. Let $\mathcal{N}_1 \subset \mathcal{N}_0$ be prime ideals of B' such that $\mathcal{N}_i \cap A = \mathcal{Q}_i$. We know by Theorem 14.26(ii), that there exists $g \in G$ such that $g(\mathcal{N}_0) = \mathcal{P}'_0$. We put $\mathcal{P}'_1 = g(\mathcal{N}_1)$. Clearly, we have $\mathcal{P}'_1 \subset \mathcal{P}'_0$ and $\mathcal{P}'_1 \cap A = \mathcal{Q}_1$, hence we are done.

14.6　Exercises

1. Let A be a Noetherian domain such that $A = \bigcap_{\mathrm{ht}(\mathcal{P})=1} A_{\mathcal{P}}$. Show that for each non-zero and non-invertible element $a \in A$, all prime ideals associated to A/aA have height one.

2. Let A be a Noetherian local domain of dimension 2. Assume that there exists $a \in A$ such that all prime ideals associated to A/aA have height one. Show that for each non-zero element $b \in A$, all prime ideals associated to A/bA have height one.

3. Show that a Noetherian domain A is a discrete valuation ring if and only if for each non-zero element $x \in K(A)$ such that $x \notin A$, then $x^{-1} \in A$.

4. Show that a Noetherian domain A is a discrete valuation ring if and only if A and $K(A)$ are the only rings containing A and contained in $K(A)$.

5. Let $F_1, F_2 \in \mathbb{C}[X_1, X_2, X_3]$ be polynomials. Put

$$X = \mathrm{Spec}_{\mathrm{m}}(\mathbb{C}[X_1, X_2, X_3]/(F_1, F_2)).$$

If $x = (a, b, c) \in X$, show that $O_{X,x}$ is a discrete valuation ring if and only if the matrix $\left(\frac{\partial F_i}{\partial X_j}(a, b, c)\right)$ has rank 2.

6. Consider the integral extension of rings

$$A = \mathbb{C}[X] \subset \mathbb{C}[X, Y]/(Y^3 + X + 1) = B.$$

Show that B is a Dedekind ring and the integral closure of A in the fraction field L of B. Find for which maximal ideal $\mathcal{M}$ of A the integral closure of $A_{\mathcal{M}}$ in L has strictly less than three maximal ideals.

7. Let V be a discrete valuation ring. Assume that the fraction field K of V has characteristic 0. Let $K \subset L$ be an algebraic field extension of degree n. Show that there are at most n discrete valution rings U such that $V \subset U$ and that L is the fraction field of U. Show that the intersection of these rings is the integral closure V' of V in L. Finally, show that V' is a principal ideal ring and that there are n different discrete valuation rings containing V if and only if a uniformizing element a of V has no quadratic factor in V'.

8. Let $P, Q \in \mathbb{C}[X, Y]$ be polynomials such that $\mathbb{C}[X, Y]$ is integral over its subring $\mathbb{C}[P, Q]$. Show that $\mathbb{C}[P, Q]$ is a polynomial ring and describe the set of prime ideals $\mathcal{P}$ of this ring such that the quotient ring $\mathbb{C}[X, Y]/\mathcal{P}\mathbb{C}[X, Y]$ is not reduced.

15

Weil divisors

Consider an affine scheme of dimension one, $X = \text{Spec}_m(R)$, in other words an affine curve. If the ring R is integrally closed, the free group generated by the points of X is the group of Weil divisors of X. Now for each point $x \in X$, the local ring $O_{X,x}$ is a discrete valution ring to which is associated a discrete valuation that we denote v_x. If $a \in K(R)$, it is easy to check that $v_x(a) = 0$ for all $x \in X$ except a finite number. Hence we can associate to a the Weil divisor $\text{div}_W(a) = \sum_{x \in X} v_x(a)[x]$. Such divisors are called principal Weil divisors; they form a subgroup of the group of Weil divisors. The quotient group is the Weil divisors class group.

This easy construction can be extended to define the group of Weil divisors and the Weil divisors class group for any integrally closed Noetherian ring. This is our first section. In the second section, we study in detail the relations between Weil divisors and reflexive rank-one modules.

Let $\mathcal{P}$ be a prime ideal of a Noetherian ring A. If $\text{ht}(\mathcal{P}) = r$, we will say that $A/\mathcal{P}$ has codimension r in A. If $\mathcal{I}$ is an ideal of A such that $A/\mathcal{P}$ has codimension r for all $\mathcal{P} \in \text{Ass}(A/\mathcal{I})$, we will say that $A/\mathcal{I}$ has pure codimension r in A.

If X is an affine scheme and $V \subset X$ a closed subvariety, the codimension of V in X will clearly be the dimension of the local ring $0_{X,V}$. We note that $\dim(V) + \text{codim}_X(V) \leq \dim(X)$ and that equality holds for all closed subvarieties of X if and only if X is equidimensional.

15.1 Weil divisors

In this chapter R is a given integrally closed Noetherian domain. We intend to study its height-one prime ideals.

Definition 15.1

(i) *To each prime ideal of height one $\mathcal{P}$ we associate an irreducible Weil divisor* $[\mathcal{P}]$ *of R.*

(ii) *The free group generated by the irreducible Weil divisors of R is the group $\operatorname{Div}_W(R)$ of Weil divisors of R.*

(iii) *A divisor $D = \sum_1^r n_i[\mathcal{P}_i]$ is effective if $n_i \geq 0$ for all i. We write in this case $D \geq 0$.*

(iv) *An effective divisor $\sum_1^r n_i[\mathcal{P}_i]$ is positive if $\sum_1^r n_i > 0$. We write in this case $D > 0$.*

Now, we want to interpret positive divisors of R as quotient rings of pure codimension one. To do this, we need the following result.

Proposition 15.2 *If $\mathcal{P}$ is a height-one prime ideal of R and $\mathcal{Q}$ a $\mathcal{P}$-primary ideal, there exists $n > 0$ such that $\mathcal{Q} = \mathcal{P}^{(n)}$.*

Proof Since $R_\mathcal{P}$ is a discrete valuation ring, any ideal in this ring is of the form $\mathcal{P}^n R_\mathcal{P}$, by Proposition 14.12. Hence $\mathcal{Q} = \mathcal{P}^n R_\mathcal{P} \cap R = \mathcal{P}^{(n)}$, for some $n > 0$. $\qquad\square$

Consider a quotient ring $R/\mathcal{I}$ of pure codimension one. By Proposition 15.2, the ideal $\mathcal{I}$ has a unique minimal primary decomposition $\mathcal{I} = \mathcal{P}_1^{(n_1)} \cap ... \cap \mathcal{P}_r^{(n_r)}$. The map $R/\mathcal{I} \to \sum_1^r n_i \mathcal{P}_i$ obviously defines a bijective correspondence between pure codimension one factors of R and effective Weil divisors. We put

$$\sum_1^r n_i[\mathcal{P}_i] = \operatorname{div}_W(R/\mathcal{I}), \quad \text{in particular} \quad [\mathcal{P}] = \operatorname{div}_W(R/\mathcal{P}).$$

If R is a Dedekind ring and $\mathcal{I}$ a non-trivial ideal of R, then $R/\mathcal{I}$ is of pure codimension one. Note that if $\operatorname{div}(R/\mathcal{I}) = \sum_1^r n_i[\mathcal{P}_i]$, then $R/\mathcal{I}$ has length $\sum_1^r n_i$. Indeed, $R/\mathcal{I} \simeq \oplus_1^r R/\mathcal{P}_i^{n_i}$ since the primary ideals $\mathcal{P}_i^{n_i}$ are pairwise comaximal. The reader will check that $R/\mathcal{P}^n$ has length n.

To each irreducible divisor $\mathcal{P}$ of R, we associate the discrete valuation $v_\mathcal{P}$, on the fraction field K of R. The discrete valuation ring of $v_\mathcal{P}$ is $R_\mathcal{P}$.

Theorem 15.3

(i) *If $x \in K^*$, there are only finitely many irreducible divisors $\mathcal{P}$ such that $v_\mathcal{P}(x) \neq 0$. If furthermore $x \in R$, then $v_\mathcal{P}(x) \geq 0$ for all $\mathcal{P}$ and*

$$xR = \bigcap_\mathcal{P} \mathcal{P}^{(v_\mathcal{P}(x))}.$$

(ii) *The map*

$$\operatorname{div}_W(.) : K^* \to \operatorname{Div}_W(R), \quad x \to \operatorname{div}_W(x) = \sum_\mathcal{P} v_\mathcal{P}(x^{-1})[\mathcal{P}],$$

is a group homomorphism.

(iii) *The divisor* div $_W(x)$ *is effective if and only if* $R \subset xR$.

(iv) *The kernel of the homomorphism* div $_W(.)$ *is the multiplicative group of units of* R.

Proof (i) Let $x = a/b$, with $a, b \in R$. Since $v_P(x) = v_P(a) - v_P(b)$, then $v_P(x) \neq 0$ implies $v_P(a) > 0$ or $v_P(b) > 0$, in other words $a \in \mathcal{P}R_P$ or $b \in \mathcal{P}R_P$. But if $a \in \mathcal{P}R_P$, the prime ideal $\mathcal{P}R_P$ of R_P is associated to aR_P, hence the prime ideal $\mathcal{P}$ of R is associated to aR. Consequently, if $v_P(x) \neq 0$ then $\mathcal{P} \in \mathrm{Ass}(R/aR) \cup \mathrm{Ass}(R/bR)$. Since both these sets are finite, we are done.

Now if $x \in R$, all prime ideals associated to xR have height one, by Theorem 14.4. Hence xR has a unique minimal primary decomposition $xR = \bigcap_i P_i^{(n_i)}$, by Proposition 15.2. This shows $v_{P_i}(x) = n_i$ on the one hand, and $v_P(x) = 0$ for $\mathcal{P} \neq P_i$ on the other.

(ii) Since $v_P((xy)^{-1}) = v_P(x^{-1}) + v_P(y^{-1})$, for all irreducible divisors $\mathcal{P}$, it is clear that div $_W(.)$ is a group homomorphism.

(iii) We have already seen that $x^{-1} \in R^*$ implies div $_W(x) \geq 0$. Conversely, if div $_W(x) \geq 0$, then $x^{-1} \in R_P$ for all height one prime ideals $\mathcal{P}$ of R. Since $R = \bigcap_{\mathrm{ht}(\mathcal{P})=1} R_P$, by Proposition 14.6, this shows $x^{-1} \in R$, hence $R \subset xR$.

(iv) Finally, div $_W(x) = 0$ if and only if $x \in R$ and $x^{-1} \in R$, and we are done. □

Remark 15.4 Note that for $a \in R$, we have div $_W(a^{-1}) = $ div $_W(R/aR)$.

Definition 15.5

(i) *The subgroup* div (K^*) *of* Div $_W(R)$ *is the group of principal Weil divisors of* R.

(ii) *We say that two Weil divisors* D *and* D' *are linearly equivalent if* $D - D'$ *is a principal Weil divisor.*

(iii) *The quotient group of* Div $_W(R)/$div (K^*) *is the Weil divisors class group* $\mathrm{Cl}_W(R)$ *of* R.

The next theorem illustrates the importance of the divisors class group.

Theorem 15.6 *The following conditions are equivalent:*

(i) *the ring* R *is a UFD;*

(ii) *all height one prime ideals are principal;*

(iii) $\mathrm{Cl}_W(R) = (0)$.

Proof Assume R is a UFD. Let $\mathcal{P}$ be a prime ideal of height one and $x \in \mathcal{P}$ a non-zero element. Since x is a product of irreducible elements, there exists an irreducible $a \in \mathcal{P}$. Since aR is a non-trivial prime ideal contained in $\mathcal{P}$, we have $aR = \mathcal{P}$, and (i) $\Rightarrow$ (ii).

Note next that (ii) and (iii) are clearly equivalent since

$$\mathrm{div}_W(x) = \mathcal{P} \iff x^{-1}R = \mathcal{P}.$$

Finally, assume (ii). Let $a \in R$ be an irreducible element and $\mathcal{P} \in \mathrm{Ass}(R/aR)$. We know, by Lemma 14.7, that $\mathrm{ht}(\mathcal{P}) = 1$. Consequently, $\mathcal{P}$ is principal, say $\mathcal{P} = bR$. Then $a \in bR$ and a irreducible imply $aR = bR$. But, by Theorem 3.8, if each irreducible element generates a prime ideal, the ring R is a UFD. □

As a special case of Theorem 15.6, we get:

Corollary 15.7 *If D is a Dedekind domain, the following conditions are equivalent:*

(i) *the ring D is a principal ideal ring;*

(ii) *all maximal ideals of D are principal;*

(iii) $\mathrm{Cl}_W(R) = (0)$.

Exercises 15.8

1. Show that the ring $R = \mathbb{C}[X,Y,Z]/(X^2+YZ)$ is integrally closed but is not a UFD. Prove $\mathrm{Cl}_W(R) \simeq \mathbb{Z}/2\mathbb{Z}$.

2. Let A be an integrally closed Noetherian domain. Show that if $\mathcal{P}$ is a height one prime ideal of A, then $\mathcal{P}A[X]$ is a height-one prime ideal of $A[X]$. Prove that the group homomorphism $\mathrm{Div}_W(A) \to \mathrm{Div}_W(A[X])$ thus defined, induces an isomorphism $\mathrm{Cl}_W(A) \simeq \mathrm{Cl}_W(A[X])$.

3. Let A be an integrally closed Noetherian domain.
   (i) If T is a finitely generated torsion A-module, we put $\mathrm{div}_W(T) = \sum_{\mathrm{ht}(\mathcal{P})=1} l(T_\mathcal{P})[\mathcal{P}]$. Show that if $0 \to T' \to T \to T'' \to 0$ is an exact sequence of finitely generated torsion A-modules, then $\mathrm{div}_W(T) = \mathrm{div}_W(T') + \mathrm{div}_W(T'')$.
   (ii) Assume there exists an exact sequence $0 \to L \xrightarrow{f} L' \to T \to 0$, where L and L' are free A-modules. Show that $\mathrm{div}_W(T) = \mathrm{div}_W(A/\det(f)) = \mathrm{div}_W(\det(f)^{-1})$.
   (iii) If M is a finitely generated A-module, show that there exists a free submodule F of M such that M/F is a torsion module and that the class of the divisor $\mathrm{div}_W(M/F)$ in $\mathrm{cl}_W(A)$ only depends on M. Denote this class by $c_1(M)$.
   (iv) Show that if $0 \to M' \to M \to M'' \to 0$ is an exact sequence of finitely generated A-modules, then $c_1(M) = c_1(M') + c_1(M'')$.

Exercise 15.9 Let $R \subset R'$ be a finite extension of integrally closed Noetherian domains.

1. If $\mathcal{P}$ is a height-one prime ideal of R, show that there are finitely many prime ideals $\mathcal{Q}$ of R', all of height one, such that $\mathcal{Q} \cap R = \mathcal{P}$.

2. If $\mathcal{Q}$ is a height-one prime ideal of R' such that $\mathcal{Q} \cap R = \mathcal{P}$, show that there exists a unique positive integer $r_{\mathcal{Q}}$ such that $\mathcal{P} R'_{\mathcal{Q}} = \mathcal{Q}^{r_{\mathcal{Q}}} R'_{\mathcal{Q}}$.

3. Consider the group homomorphism $\pi : \mathrm{Div}_W(R) \to \mathrm{Div}_W(R')$ defined by
$$\pi([\mathcal{P}]) = \sum_{\mathcal{Q} \cap R = \mathcal{P}} r_{\mathcal{Q}}[\mathcal{Q}].$$
If $x \in K(R)^* \subset K(R')^*$, denote by $\mathrm{div}_{WR}(x)$ and $\mathrm{div}_{WR'}(x)$ the principal divisors of $\mathrm{div}_W(R)$ and $\mathrm{div}_W(R')$ defined by x and show that $\pi(\mathrm{div}_{WR}(x)) = \mathrm{div}_{WR'}(x)$

4. Show that there exists a unique group homomorphism $\psi : \mathrm{Cl}_W(R) \to \mathrm{Cl}_W(R')$ such that the following diagram is commutative

$$
\begin{array}{ccc}
K(R)^* & \subset & K(R')^* \\
\downarrow & & \downarrow \\
\mathrm{Div}_W(R) & \xrightarrow{\pi} & \mathrm{Div}_W(R') \\
\downarrow & & \downarrow \\
\mathrm{Cl}_W(R) & \xrightarrow{\psi} & \mathrm{Cl}_W(R').
\end{array}
$$

15.2 Reflexive rank-one modules and Weil divisors

We recall that R is again a given integrally closed Noetherian domain.

Consider a domain A. We recall that an A-module M is torsion-free if for all non-zero elements $a \in A$ and $x \in M$ we have $ax \neq 0$. In this case, for each multiplicatively closed part S of A, the localization homomorphism $M \to S^{-1}M$ is injective; furthermore, if $T = A \setminus \{0\}$, there is a natural inclusion $S^{-1}M \subset T^{-1}M$. Note that

$$M = \bigcap_{\mathcal{P} \in \mathrm{Spec}(A)} M_{\mathcal{P}};$$

the proof is easy and left to the reader.

Definition 15.10 *Let A be a domain, $T = A \setminus \{0\}$ and N an A-module. We define the rank of N as the rank of the $T^{-1}A$-vector space $T^{-1}N$.*

Note that if $\mathrm{rk}(N) = r$ and $\mathrm{rk}(N') = r'$, then $\mathrm{rk}(N \otimes_A N') = rr'$. Indeed we recall that $T^{-1}(N \otimes_A N') \simeq T^{-1}N \otimes_{T^{-1}A} T^{-1}N'$.

Proposition 15.11 *If M is an R-module, the dual $M^{\check{}}$ of M is torsion-free and satisfies*

$$M^{\check{}} = \bigcap_{\mathrm{ht}(\mathcal{P})=1} (M^{\check{}})_{\mathcal{P}}.$$

Proof Consider $f \in M^{\check{}}$ and $a \in R$. If $af = 0$, then $af(x) = 0$ for all $x \in M$. But this implies $f(x) = 0$ for all $x \in M$, hence $f = 0$ and $M^{\check{}}$ is torsion-free. Note that there is a natural identification $(M^{\check{}})_{\mathcal{P}} = M_{\tilde{\mathcal{P}}}^{\check{}}$.

The inclusion $M^{\check{}} \subset \bigcap_{\mathrm{ht}(\mathcal{P})=1} M_{\tilde{\mathcal{P}}}^{\check{}}$ is obvious. Now, if $g \in \bigcap_{\mathrm{ht}(\mathcal{P})=1} M_{\tilde{\mathcal{P}}}^{\check{}}$ we have

$$g(M) \subset \bigcap_{\mathrm{ht}(\mathcal{P})=1} R_{\mathcal{P}},$$

but this intersection is R by Proposition 14.6, and $g \in M^{\check{}}$. □

Corollary 15.12 *A finitely generated R-module M is reflexive if and only if it is torsion-free and*

$$M = \bigcap_{\mathrm{ht}(\mathcal{P})=1} M_{\mathcal{P}}.$$

Proof Note first that, by Proposition 15.11, a reflexive module is torsion-free. Hence we can assume that M is torsion-free.

Let $\mathcal{P}$ be a prime ideal of height-one. Since $M_{\mathcal{P}}$ is a torsion-free finitely generated module on the principal ideal ring $A_{\mathcal{P}}$, it is free, hence reflexive. Consider then the natural commutative diagram

$$\begin{array}{ccc} M & \to & M^{\check{}\check{}} \\ \downarrow & & \downarrow \\ M_{\mathcal{P}} & \simeq & M_{\tilde{\mathcal{P}}}^{\check{}\check{}}. \end{array}$$

Since $M^{\check{}\check{}} = \bigcap_{\mathrm{ht}(\mathcal{P})=1} M_{\tilde{\mathcal{P}}}^{\check{}\check{}}$, we find $M \simeq M^{\check{}\check{}} \Longleftrightarrow M = \bigcap_{\mathrm{ht}(\mathcal{P})=1} M_{\mathcal{P}}$. □

Corollary 15.13 *A finitely generated torsion-free R-module M is reflexive if and only if for all non-zero elements $a \in R$, all prime ideals $\mathcal{P} \in \mathrm{Ass}(M/aM)$ are of height one.*

Proof First we assume that M is reflexive. Let $\mathcal{Q}$ be a prime ideal of height at least two. Asumme that $x \in M$ is such that $x\mathcal{Q} \subset aM$. This shows $xa^{-1} \in M_{\mathcal{P}}$ for all height-one prime ideals $\mathcal{P}$. By Corollary 15.12, this proves $xa^{-1} \in M$, hence $x \in aM$. Consequently $\mathcal{Q} \notin \mathrm{Ass}(M/aM)$ is impossible.

To prove the converse, we apply Corollary 15.12 once more. Consider $x \in \bigcap_{\mathrm{ht}(\mathcal{P})=1} M_{\mathcal{P}}$. We want to prove $x \in M$. Let $y \in M$ and $a \in R \setminus \{0\}$ be such that $x = y/a$. We have $y/a \in M_{\mathcal{P}}$, hence $y \in aM_{\mathcal{P}}$, for all height-one prime ideals $\mathcal{P}$. This shows that $aM : y$ is not contained in any height-one prime ideal. If $x \notin M$, then $y \notin aM$ and $\mathrm{cl}(y) \neq 0 \in M/aM$. In this case $(0) : \mathrm{cl}(y) = (aM : y)$ is contained in a prime ideal associated to M/aM. Since they are all of height one, this is a contradiction, and $x \in M$. $\qquad\square$

Applying Theorem 15.11 and Corollary 15.12, we find:

Corollary 15.14 *If M is a finitely generated R-module, then $M\check{\ }$ is reflexive.*

Next we note the following obvious but convenient consequence of Corollary 15.12:

Corollary 15.15 *A homomorphism $f : M \to N$ of reflexive finitely generated R-modules is injective (resp. surjective, bijective) if and only if the homomorphisms $f_{\mathcal{P}} : M_{\mathcal{P}} \to N_{\mathcal{P}}$ are injective (resp. surjective, bijective) for all prime ideals $\mathcal{P}$ of height-one.*

In the remainder of this section we are particularly interested in finitely generated rank-one, reflexive modules.

Proposition 15.16

(i) *If M and N are finitely generated rank-one, reflexive R-modules, then so is $(M \otimes_R N)\check{\ }\check{\ }$.*

(ii) *If M is a finitely generated rank-one, reflexive R-module, then*

$$(M\check{\ } \otimes_R M)\check{\ }\check{\ } \simeq R.$$

(iii) *The operation described in (i) gives to the set of all isomorphism classes of finitely generated rank-one, reflexive R-modules the structure of a commutative group.*

Proof (i) We note first that the dual of a finitely generated module M over a Noetherian ring A is finitely generated. Indeed, a surjective homomorphism $nA \to M$ induces an injective homomorphism $M\check{\ } \to (nA)\check{\ } \simeq nA$. So $M\check{\ }$ is isomorphic to a submodule of a finitely generated A-module and is therefore finitely generated.

By Proposition 15.11, $(M \otimes_R N)\check{\ }\check{\ }$ is torsion-free and reflexive. We must show that it has rank one. We put $T = R \setminus \{0\}$ and recall the isomorphism $T^{-1}(M \otimes_R N) \simeq T^{-1}M \otimes_{T^{-1}R} T^{-1}N$. It shows that the rank of the vector space $T^{-1}(M \otimes_R N)$ is one. Now, by Proposition 7.24, we have

$$T^{-1}((M \otimes_R N)\check{\ }\check{\ }) \simeq (T^{-1}(M \otimes_R N))\check{\ }\check{\ }.$$

This proves $\operatorname{rk}((M \otimes_R N)^{\check{}\check{}}) = 1$.

(ii) Consider the homomorphism

$$(M^{\check{}} \otimes_R M) \to R, \quad f \otimes_R x \to f(x).$$

It induces a homomorphism $(M^{\check{}} \otimes_R M)^{\check{}\check{}} \to R^{\check{}\check{}} \simeq R$.

If $\mathcal{P}$ is a height-one prime ideal, then $M_{\mathcal{P}}$ is a rank-one free $R_{\mathcal{P}}$-module, hence the induced homomorphism $(M_{\mathcal{P}}^{\check{}} \otimes_R M_{\mathcal{P}})^{\check{}\check{}} \to R_{\mathcal{P}}$ is an isomorphism. We are done by Corollary 15.15.

To prove (iii), note that the natural isomorphisms

$$(M \otimes_R N) \otimes_R P \simeq M \otimes_R (N \otimes_R P),$$

$$(M \otimes_R N) \simeq (N \otimes_R M) \quad \text{and} \quad R \otimes_R N \simeq N$$

show immediately that the operation is associative and commutative and that R is a neutral element. Since by (ii) every element has an inverse, we are done. ☐

From now on, we shall say quasi-invertible R-module for finitely generated rank-one, reflexive R-module. We go on studying the group of isomorphism classes of quasi-invertible R-modules.

We have chosen a description of the group operation that has advantages but which is not always the most convenient. The following lemma shows that a different approach is possible.

Lemma 15.17 *Let M and N be quasi-invertible R-modules.*

(i) *The R-module $\operatorname{Hom}_R(M, N)$ is quasi-invertible.*

(ii) *There is a natural isomorphism $(M \otimes_R N)^{\check{}\check{}} \simeq \operatorname{Hom}_R(M^{\check{}}, N)$.*

Proof (i) Clearly, $\operatorname{Hom}_R(M, N)$ is a rank-one finitely generated R-module. If $\mathcal{P}$ is a height-one prime ideal, there is a natural identification

$$(\operatorname{Hom}_R(M, N))_{\mathcal{P}} = \operatorname{Hom}_{R_{\mathcal{P}}}(M_{\mathcal{P}}, N_{\mathcal{P}}),$$

by Proposition 7.24. Now if $g \in \bigcap_{\operatorname{ht}(\mathcal{P})=1} \operatorname{Hom}_{R_{\mathcal{P}}}(M_{\mathcal{P}}, N_{\mathcal{P}})$, then $g(M) \subset \bigcap_{\operatorname{ht}(\mathcal{P})=1} N_{\mathcal{P}} = N$, hence $g \in \operatorname{Hom}_R(M, N)$. This shows

$$\operatorname{Hom}_R(M, N) = \bigcap_{\operatorname{ht}(\mathcal{P})=1} \operatorname{Hom}_{R_{\mathcal{P}}}(M_{\mathcal{P}}, N_{\mathcal{P}}) = \bigcap_{\operatorname{ht}(\mathcal{P})=1} (\operatorname{Hom}_R(M, N))_{\mathcal{P}},$$

and $\operatorname{Hom}_R(M, N)$ is reflexive by Corollary 15.12.

(ii) Consider the natural homomorphism $\pi : (M \otimes_R N) \to \mathrm{Hom}_R(M\check{\ }, N)$, defined by $\pi(x \otimes_R y)(f) = f(x)y$. If M and N are free, π is obviously an isomorphism. Since $M_\mathcal{P}$ and $N_\mathcal{P}$ are free $R_\mathcal{P}$-modules for all height-one prime ideals $\mathcal{P}$, this shows that $\pi_\mathcal{P}$ is an isomorphism for all height-one prime ideals $\mathcal{P}$. By Corollary 15.15, $\pi\check{\ }\check{\ }: (M \otimes_R N)\check{\ }\check{\ } \to (\mathrm{Hom}_R(M\check{\ }, N))\check{\ }\check{\ }$ is an isomorphism. But $\mathrm{Hom}_R(M\check{\ }\check{\ }, N)$ is reflexive and we are done. $\qquad \square$

From now on, in this section, we are particularly interested in the quasi-invertible submodules of the fraction field of R.

If $\mathcal{F}$ and $\mathcal{G}$ are two non-zero submodules of the fraction field K of R, we put

$$\mathcal{F} :_K \mathcal{G} = \{a \in K,\ a\mathcal{G} \subset \mathcal{F}\}.$$

Note that if $\mathcal{F}$ and $\mathcal{G}$ are finitely generated, so is $\mathcal{F} :_K \mathcal{G}$. Indeed, if $b \in \mathcal{G}$ is non-zero and $(a_1, \dots, a_k)$ is a system of generators of $\mathcal{F}$, then $\mathcal{F} :_K \mathcal{G}$ is a submodule of the module finitely generated by $(a_1/b, \dots, a_k/b)$.

Proposition 15.18

(i) *A non-zero proper ideal $\mathcal{I} \subset R$ is reflexive if and only if all prime ideals associated to $\mathcal{I}$ have height one.*

(ii) *If $\mathcal{I}$ and $\mathcal{J}$ are reflexive non-zero ideals of R, then $\mathcal{J} :_K \mathcal{I}$ is reflexive and the natural homomorpism $\mathcal{J} :_K \mathcal{I} \to \mathrm{Hom}_R(\mathcal{I}, \mathcal{J})$ is an isomorphism.*

Proof (i) We know that $\mathcal{I} \subset \bigcap_{\mathrm{ht}(\mathcal{P})=1} \mathcal{I}_\mathcal{P}$ and that equality holds if and only if $\mathcal{I}$ is reflexive.

Now clearly,

$$a \in \bigcap_{\mathrm{ht}(\mathcal{P})=1} \mathcal{I}_\mathcal{P} \iff (\mathcal{I} : a) \not\subset \mathcal{P} \quad \text{for} \quad \mathrm{ht}(\mathcal{P}) = 1.$$

Assume first that $\mathcal{I}$ is reflexive. If $\mathcal{Q} \in \mathrm{Ass}(R/\mathcal{I})$, there exists $a \notin \mathcal{I}$ such that $(\mathcal{I} : a) = \mathcal{Q}$. This shows that there exists a prime ideal of height one $\mathcal{P}$ such that $\mathcal{Q} \subset \mathcal{P}$, hence $\mathcal{Q} = \mathcal{P}$.

Conversely, if $\mathcal{I}$ is not reflexive, there exists

$$a \notin \mathcal{I}, \quad \text{with} \quad a \in \bigcap_{\mathrm{ht}(\mathcal{P})=1} \mathcal{I}_\mathcal{P} \subset \bigcap_{\mathrm{ht}(\mathcal{P})=1} R_\mathcal{P} = R.$$

This shows $(\mathcal{I} : a) \not\subset \mathcal{P}$ for all prime ideals $\mathcal{P}$ of height one. But there exists $\mathcal{Q} \in \mathrm{Ass}(R/\mathcal{I})$ such that $(\mathcal{I} : a) \subset \mathcal{Q}$. Obviously, $\mathrm{ht}(\mathcal{Q}) > 1$.

(ii) First we use Corollary 15.12 to show that $\mathcal{J} :_K \mathcal{I}$ is reflexive. Indeed, if

$$x \in \bigcap_{\mathrm{ht}(\mathcal{P})=1} (\mathcal{J}_\mathcal{P} :_K \mathcal{I}_\mathcal{P}), \quad \text{then} \quad x\mathcal{I} \subset \bigcap_{\mathrm{ht}(\mathcal{P})=1} \mathcal{J}_\mathcal{P} = \mathcal{J} \quad \text{and} \quad x \in (\mathcal{J} :_K \mathcal{I}).$$

Next we consider the natural inclusion of finitely generated rank-one, reflexive R-modules $\mathcal{J} :_K \mathcal{I} \subset \mathrm{Hom}_R(\mathcal{I}, \mathcal{J})$. If $\mathcal{P}$ is a prime ideal of height-one, then $\mathcal{I}_\mathcal{P}$ and $\mathcal{J}_\mathcal{P}$ are principal. This implies $(\mathcal{J}_\mathcal{P} :_K \mathcal{I}_\mathcal{P}) = \mathrm{Hom}_{R_\mathcal{P}}(\mathcal{I}_\mathcal{P}, \mathcal{J}_\mathcal{P})$, hence $(\mathcal{J} :_K \mathcal{I}) = \mathrm{Hom}_R(\mathcal{I}, \mathcal{J})$, by Corollary 15.15. $\qquad \square$

The advantages of our next lemma will appear very soon, so don't be put off by its appearance.

Lemma 15.19 *If $\mathcal{P}_i$, with $i = 1, ..., r$, are height-one prime ideals of R and $l_i, n_i, m_i \geq 0$ for $i = 1, ..., r$, then*

$$\left(\bigcap_{i=1}^{r} \mathcal{P}_i^{(n_i)}\right) :_K \left(\bigcap_{i=1}^{r} \mathcal{P}^{(m_i)}\right) = \left(\bigcap_{i=1}^{r} \mathcal{P}_i^{(n_i+l_i)}\right) :_K \left(\bigcap_{i=1}^{r} \mathcal{P}^{(m_i+l_i)}\right).$$

Proof Put

$$\mathcal{F} = \left(\bigcap_{i=1}^{r} \mathcal{P}_i^{(n_i)}\right) :_K \left(\bigcap_{i=1}^{r} \mathcal{P}^{(m_i)}\right)$$

and

$$\mathcal{G} = \left(\bigcap_{i=1}^{r} \mathcal{P}_i^{(n_i+l_i)}\right) :_K \left(\bigcap_{i=1}^{r} \mathcal{P}^{(m_i+l_i)}\right).$$

Note first that $\mathcal{F}$ and $\mathcal{G}$ are both finitely generated submodules of the fraction field K of R. Since they are reflexive by Proposition 15.18, we have

$$\mathcal{F} = \bigcap_{\mathrm{ht}(\mathcal{P})=1} \mathcal{F}_\mathcal{P} \quad \text{and} \quad \mathcal{G} = \bigcap_{\mathrm{ht}(\mathcal{P})=1} \mathcal{G}_\mathcal{P}.$$

Consequently it suffices to show that $\mathcal{F}_\mathcal{P} = \mathcal{G}_\mathcal{P}$ for each prime ideal $\mathcal{P}$ of height one.

If $\mathcal{P} \neq \mathcal{P}_i$ for $i : 1, ..., r$, then $\mathcal{F}_\mathcal{P} = \mathcal{G}_\mathcal{P} = R_\mathcal{P}$.

If $\mathcal{P} = \mathcal{P}_i$, let u be a uniformizing element in the ring $R_{\mathcal{P}_i}$. Then

$$\mathcal{F}_{\mathcal{P}_i} = \mathcal{P}_i^{(n_i)} R_{\mathcal{P}_i} :_K \mathcal{P}_i^{(m_i)} R_{\mathcal{P}_i} = u^{n_i} R_{\mathcal{P}_i} :_K u^{m_i} R_{\mathcal{P}_i} = u^{n_i - m_i} R_{\mathcal{P}_i}.$$

This obviously shows our technical lemma. $\qquad \square$

As an immediate consequence of it, we get the following useful corollary.

Corollary 15.20 *Let* $\mathcal{P}_i$, *with* $i = 1, ..., r$, *be height-one prime ideals of* R *and* n_i, m_i *positive integers. If* $\mathcal{I} = \bigcap_{i=1}^{r} \mathcal{P}_i^{(n_i)}$ *and* $\mathcal{J} = \bigcap_{i=1}^{r} \mathcal{P}_i^{(m_i)}$, *the map which associates to the Weil divisor* $D = \sum_{i=1}^{r}(m_i - n_i)[\mathcal{P}_i]$ *the quasi-invertible* R-submodule $\mathcal{I} :_K \mathcal{J}$ *is well defined.*
  We denote this quasi-invertible module by $O(D)$.

We note that $O(D)$ is a submodule of K.

Theorem 15.21 *If* π_W *is the map from the group* $\mathrm{Div}_W(R)$ *to the group of isomorphism classes of quasi-invertible* R-modules *which associates to a divisor* $\Delta = \sum_{i=1}^{r}(m_i - n_i)[\mathcal{P}_i]$, *with* $n_i, m_i \geq 0$, *the isomorphism class of* $O(\Delta) = (\bigcap_{i=1}^{r} \mathcal{P}_i^{(n_i)}) :_K (\bigcap_{i=1}^{r} \mathcal{P}_i^{(m_i)})$, *then:*

(i) *the map* π_W *is a surjective group homomorphism;*

(ii) *the kernel of* π_W *is the group of principal divisors and so* π_W *induces an isomorphism between* $\mathrm{Cl}_W(R)$ *and the group of isomorphism classes of quasi-invertible* R-modules.

Proof We denote by $[L]$ the isomorphism class of a quasi-invertible R-module L. We consider the operation in the group as a product. In other words we write

$$[L][L'] = [(L \otimes_R L')^{\vee\vee}] = [\mathrm{Hom}_R(L^{\vee}, L')] = [\mathrm{Hom}_R(L'^{\vee}, L)] \text{ and } [L]^{-1} = [L^{\vee}].$$

We recall that π_W is well defined by Corollary 15.20. First we show that this map is a group homomorphism. We put $\mathcal{I} = \bigcap_{i=1}^{r} \mathcal{P}_i^{(n_i)}$ and $\mathcal{J} = \bigcap_{i=1}^{r} \mathcal{P}_i^{(m_i)}$. Consider the effective divisor $D = \sum_{i=1}^{r} m_i[\mathcal{P}_i]$. Then

$$\pi_W(D) = [R :_K \mathcal{J}] = [\mathcal{J}^{\vee}].$$

As a special case we get $\pi_W(0) = [R]$. Put $D' = \sum_{i=1}^{r} n_i[\mathcal{P}_i]$. Then

$$\pi_W(-D') = [\mathcal{I} :_K R] = [\mathcal{I}] = [\mathcal{I}^{\vee}]^{-1} = \pi_W(D')^{-1}.$$

Recalling the isomorphisms

$$\mathcal{I} :_K \mathcal{J} \simeq \mathrm{Hom}_R(\mathcal{J}, \mathcal{I}) \simeq (\mathcal{J}^{\vee} \otimes_R \mathcal{I})^{\vee\vee},$$

we have proved

$$\pi_W(D - D') = [\mathrm{Hom}_R(\mathcal{J}, \mathcal{I})] = [\mathcal{J}^{\vee} \otimes_R \mathcal{I})^{\vee\vee}] = \pi_W(D)\pi_W(D')^{-1}.$$

Next we want to prove $\pi_W(D + D') = \pi_W(D)\pi_W(D')$ or equivalently

$$\pi_W(-D - D') = \pi_W(D)^{-1}\pi_W(D')^{-1}.$$

In order to do so, we must show

$$(\mathcal{I} \otimes_R \mathcal{J})^{\vee\vee} \simeq \bigcap_i \mathcal{P}_i^{(n_i+m_i)}.$$

This can be done by using Corollary 15.15 once more. Indeed, there are obvious natural homomorphisms

$$(\mathcal{I} \otimes_R \mathcal{J}) \to \mathcal{I}\mathcal{J} \quad \text{and} \quad \mathcal{I}\mathcal{J} = (\bigcap_i \mathcal{P}_i^{(n_i)})(\bigcap_i \mathcal{P}_i^{(m_i)}) \to \bigcap_i \mathcal{P}_i^{(n_i+m_i)}$$

and it suffices to prove that for any height-one prime ideal $\mathcal{P}$ of R, they induce isomorphisms

$$(\mathcal{I}_\mathcal{P} \otimes_{R_\mathcal{P}} \mathcal{J}_\mathcal{P}) \simeq \mathcal{I}_\mathcal{P}\mathcal{J}_\mathcal{P} \quad \text{and} \quad \mathcal{I}_\mathcal{P}\mathcal{J}_\mathcal{P} \simeq \bigcap_i(\mathcal{P}_i^{(n_i+m_i)}R_\mathcal{P}).$$

Let $u \in R_\mathcal{P}$ be a uniformizing element of this discrete valuation ring. There exist integers $n \geq 0$ and $m \geq 0$ such that $\mathcal{I}_\mathcal{P} = u^n R_\mathcal{P}$ and $\mathcal{J}_\mathcal{P} = u^m R_\mathcal{P}$ and it is clear that the natural homomorphism $u^n R_\mathcal{P} \otimes_{R_\mathcal{P}} u^m R_\mathcal{P} \to u^{n+m} R_\mathcal{P}$ is an isomorphism. Now if $\mathcal{P} \neq \mathcal{P}_i$ for all i, then

$$n = m = 0 \quad \text{and} \quad \bigcap_i(\mathcal{P}_i^{(n_i+m_i)}R_\mathcal{P}) = R_\mathcal{P}.$$

If $\mathcal{P} = \mathcal{P}_t$, then

$$n = n_t, \quad m = m_t \quad \text{and} \quad \bigcap_i(\mathcal{P}_i^{(n_i+m_i)}R_\mathcal{P}) = u^{n_t+m_t}R_\mathcal{P}$$

and we are done.

Now if Δ_1 and Δ_2 are divisors, there exist effective divisors D_i and D_i', with $i = 1, 2$, such that $\Delta_i = D_i - D_i'$. We have

$$\begin{aligned}
\pi_W(\Delta_1 - \Delta_2) &= \pi_W(D_1 + D_2' - (D_1' + D_2)) = \pi_W(D_1 + D_2')\pi_W(D_1' + D_2)^{-1} \\
&= \pi_W(D_1)\pi_W(D_2')\pi_W(D_1')^{-1}\pi_W(D_2)^{-1} \\
&= \pi_W(D_1 - D_1')\pi_W(D_2 - D_2')^{-1} = \pi_W(\Delta_1)\pi_W(\Delta_2)^{-1}.
\end{aligned}$$

We have proved that π_W is a group homomorphism.

Proving that π_W is surjective is easier. Let $\mathcal{M}$ be a quasi-invertible R-module. If $T = R \setminus \{0\}$, consider an isomorphism $\psi : T^{-1}\mathcal{M} \simeq T^{-1}R$. Since $\psi(\mathcal{M}) \subset T^{-1}R$ is a finitely generated R-module, there exists $t \in T$ such that $t\psi(\mathcal{M}) \subset R$. Note that $\mathcal{M} \simeq t\psi(\mathcal{M})$. Hence the ideal $\mathcal{I} = t\psi(\mathcal{M})$ is reflexive. If $D = \operatorname{div}_W(R/\mathcal{I})$, we have proved that $[\mathcal{M}] = \pi_W(-D)$.

To conclude, we must prove that the kernel of π_W is the group of principal divisors. As before put $\mathcal{I} = \bigcap_{i=1}^r \mathcal{P}_i^{(n_i)}$ and $\mathcal{J} = \bigcap_{i=1}^r \mathcal{P}_i^{(m_i)}$, with $n_i, m_i \geq 0$. If $D = \sum_{i=1}^r (m_i - n_i)[\mathcal{P}_i]$, then

$$D \in \ker \pi_W \iff [O(D)] = [R] \iff \mathcal{I} :_K \mathcal{J} \simeq R.$$

But this is equivalent to the existence of an element $x \in K(R)$ such that $(\mathcal{I} :_K \mathcal{J}) = xR$. This last relation is equivalent with

$$\mathcal{I}_{\mathcal{P}} = x\mathcal{J}_{\mathcal{P}} \quad \text{for all} \quad \mathcal{P} \quad \text{such that} \quad \text{ht}(\mathcal{P}) = 1.$$

But this means $v_{\mathcal{P}_i}(x^{-1}) = m_i - n_i$ for each i and $v_{\mathcal{P}}(x) = 0$ if $\mathcal{P} \neq \mathcal{P}_i$ for all i. In other words $D \in \ker \pi$ if and only if there exists an element $x \in K(R)^*$ such that $\text{div}_W(x) = D$. $\qquad\square$

Corollary 15.22 *An integrally closed Noetherian domain is a UFD if and only if all quasi-invertible R-modules are free.*

Proof The ring R is a UFD if and only if $\text{Cl}_W(R) = (0)$, by Theorem 15.6. But, by Theorem 15.21, we find that $\text{Cl}_W(R) = (0)$ if and only if each quasi-inverible module is isomorphic to R. $\qquad\square$

Exercise 15.23 Let $R \subset R'$ be a finite extension of integrally closed Noetherian rings.

1. If M is a quasi-invertible R-module, show that $(M \otimes_R R')^{\vee\vee}$ is a quasi-invertible R'-module.

2. Show that the map $\psi : \text{cl}_W(R) \to \text{cl}_W(R')$ defined by $\phi(M) = (M \otimes_R R')^{\vee\vee}$ is a group homomorphism.

3. Compare the homomorphism ϕ with the homomorphism ψ defined in Exercise 15.9 (4).

15.3 Exercises

The ring R is a given integrally closed Noetherian domain.

1. Let $\mathcal{I} \subset R$ be an ideal. Show that the natural injective homomorphism $\mathcal{I}^{\vee\vee} \to R^{\vee\vee} \simeq R$ is an isomorphism if and only if $\mathcal{I}$ is not contained in any height one prime ideal of R.

2. Let $\mathcal{I} \subset R$ be a reflexive ideal. If $a \in \mathcal{I}$ is a non-zero element, show $(a) : \mathcal{I} \simeq \mathcal{I}^{\vee}$.

3. Let $D = \sum_{i=1}^{r} n_i \mathcal{P}_i$ be a positive Weil divisor and $\mathcal{I} = \bigcap_{i=1}^{r} \mathcal{P}_i^{n_i}$. The natural map $\mathcal{I} \subset R$ induces an injective homomorphism $R \subset \mathcal{I}^{\vee}$ (the image of 1 is the natural map). Show that $\text{Ass}(\mathcal{I}^{\vee}/R) = \{\mathcal{P}_1, \ldots, \mathcal{P}_r\}$ and that $(\mathcal{I}^{\vee}/R)_{\mathcal{P}_i}$ is an $R_{\mathcal{P}_i}$-module of length n_i.

4. Let $\mathcal{I} \subset R$ be a reflexive ideal. Assume that $\mathcal{I}$ is generated by two elements a and b. Show that the kernel of the surjective homomorphism $2R \xrightarrow{(a,b)} \mathcal{I}$ is isomorphic to $\mathcal{I}$.

5. Show that a finitely generated R-module M is reflexive if and only if M is the kernel of a homomorphism of finitely generated free R-modules.

6. If $\mathcal{A}$ and $\mathcal{B}$ are two quasi-invertible submodules of the fraction field of R, show that $\mathcal{A} \simeq \mathcal{B}$ if and only if there exist non-zero elements $a, b \in R$ such that $b\mathcal{A} = a\mathcal{B}$.

7. Assume that for each $c \in \mathrm{cl}_W(R)$ there exists an integer n such that $nc = 0$. Show that a prime ideal of height one of R is the radical of a principal ideal.

8. Assume that the class of the divisor $D = \sum_{i=1}^{r} n_i \mathcal{P}_i$ generates $\mathrm{cl}_W(R)$. Show that for each quasi-invertible ideal $\mathcal{I}$ there exist positive integers s_i, t_i, for $i = 1, \ldots, r$ and elements $a, b \in R$ such that

$$a(\mathcal{I} \cap \mathcal{P}_1^{s_1} \cap \ldots \cap \mathcal{P}_r^{s_r}) = b(\mathcal{P}_1^{t_1} \cap \ldots \cap \mathcal{P}_r^{t_r}).$$

16

Cartier divisors

If Weil divisors are often convenient to study curves and surfaces, they do not behave well enough in higher dimensional geometry. They do not behave well under base change either: more precisely if $R \to R'$ is a homomorphism of integrally closed Noetherian rings, we meet real difficulties in trying to construct group homomorphisms $\operatorname{div}_W(R) \to \operatorname{div}_W(R')$ and $\operatorname{cl}_W(R) \to \operatorname{cl}_W(R')$. When the local rings $R_\mathcal{M}$ are UFDs for all maximal ideals $\mathcal{M}$, many of these drawbacks disappear. This is precisely when all Weil divisors become Cartier divisors.

We present and describe Cartier divisors in our first section. They are defined for any Noetherian ring. When the ring is integrally closed, they form a subgroup of the group of Weil divisors.

In the second section of this chapter, we study the relations between Cartier divisors and finitely generated rank-one locally free modules. Such modules are called invertible. The isomorphism classes of invertible modules form a group, the Picard group of the ring. When the ring is integrally closed, the Picard group is a subgroup of the group of isomorphism classes of quasi-invertible modules.

16.1   Cartier divisors

Definition 16.1 *A non-zero divisor element of a ring A is also called a regular element.*

If $T \subset A$ is the multiplicatively closed part formed by all regular elements of A, the fraction ring $T^{-1}A$ is the total fraction ring $T(A)$ of A.

Note that a regular element of $T(A)$ is invertible. Indeed, if $a/b \in T(A)$ is regular, then a is regular in A and $b/a \in T(A)$. Hence the set of regular elements of $T(A)$ is the group $U(T(A))$ of units of $T(A)$. The units $U(A)$ of A form a subgroup of this group.

Definition 16.2

(i) *The quotient group $U(T(A))/U(A)$ is called the group of principal Cartier divisors of A.*

(ii) *The class in $U(T(A))/U(A)$ of a regular element $x \in T(A)$ is denoted by $\operatorname{div}_C(x)$.*

Note that when A is a domain, then the group of principal Cartier divisors is $K(A)^*/U(A)$.

In the group of principal Cartier divisors we use an additive notation: we write

$$\operatorname{div}_C(xy) = \operatorname{div}_C(x) + \operatorname{div}_C(y), \quad \text{hence} \quad \operatorname{div}_C(x/y) = \operatorname{div}_C(x) - \operatorname{div}_C(y).$$

Next we study the relations between principal Cartier divisors of A and free A-submodules of $T(A)$.

Proposition 16.3 *The correspondence $\operatorname{div}_C(x) \longleftrightarrow xA$, between principal Cartier divisors of A and free A-submodules of $T(A)$ is bijective.*

Proof If $x \in T(A)$ is a regular element, it is clear that xA is a free A-module. Conversely, let $xA \subset T(A)$ be a free A-module. Put $x = a/t$, with $t \in T$. Then $ab = 0$, with $b \in A$, implies $xb = 0$ hence $b = 0$. This shows that a is regular in A, hence that $x \in U(T(A))$. To conclude, note that if $x, y \in U(T(A))$, then

$$xA = yA \iff x/y \in U(A) \iff \operatorname{div}_C(x) = \operatorname{div}_C(y).$$

$\square$

Definition 16.4

(i) *A principal Cartier divisor $\operatorname{div}_C(x)$ is effective if $A \subset xA$ or equivalently if $x^{-1} \in A$.*

(ii) *An effective principal Cartier divisor $\operatorname{div}_C(x)$ is positive if $x \notin A$.*

(iii) *We write $\operatorname{div}_C(x) \geq \operatorname{div}_C(y)$ if the principal Cartier divisor $\operatorname{div}_C(x/y)$ is effective, i.e. if $y/x \in A$.*

Note the obvious following result:

Proposition 16.5 *If $\operatorname{div}_C(x)$ and $\operatorname{div}_C(y)$ are effective principal Cartier divisors, so is $\operatorname{div}_C(xy)$. If furthermore $\operatorname{div}_C(x)$ or $\operatorname{div}_C(y)$ is positive, then so is $\operatorname{div}_C(xy)$.*

Now, we want to define the group of Cartier divisors of any Noetherian ring A. Before we do so, we recall that a principal Cartier divisor is a free A-submodule of $T(A)$. We are tempted to define Cartier divisors as finitely generated locally free A-submodules of $T(A)$. Unfortunatly, this is not always convenient; the total fraction ring may be much too small. When A is a domain, this approach is excellent.

Exercise 16.6 Let A be a Noetherian domain and K its fraction field. If M and N are submodules of K we denote by MN the submodule of K generated by all xy, with $x \in M$ and $y \in N$.

1. Show that if $\mathcal{F}$ and $\mathcal{F}'$ are finitely generated locally free, rank-one A-submodules of K, then so is $\mathcal{F}\mathcal{F}'$.

2. Show that the set of finitely generated locally free, rank-one A-submodules of K, equipped with this product, is a commutative group with neutral element A itself.
   This group, denoted by $\operatorname{Div}_C(A)$, is called the group of Cartier divisors of A.

3. Show that $\operatorname{Div}_C(A)$ is ordered by $\mathcal{F} \geq 0 \Longleftrightarrow A \subset \mathcal{F}$.

4. Show that the map $\pi : K^* \to \operatorname{div}_C(A)$ defined by $\pi(x) = x^{-1}A$ is a group homomorphism whose kernel is the group of units of A and whose image is the group of free rank-one A-submodules of K (also called the group of principal Cartier divisors of A).
   The group $\operatorname{coker} \pi$ is denoted by $\operatorname{Cl}_C(A)$ and called the group of classes of Cartier divisors of A.

When the ring A is not a domain, in other words when it has no fraction field, we cannot define and study its Cartier divisors so pleasantly. So be patient, while we develop the necessary machinery!

Definition 16.7 *Let A be a Noetherian ring and $\mathcal{I} \subset A$ be a locally free ideal of A.*
If $\mathcal{I}$ is a proper ideal, the factor ring $A/\mathcal{I}$ is called a positive Cartier divisor. We denote this positive Cartier divisor by $\operatorname{div}_C(A/\mathcal{I})$.
If $\mathcal{I} = A$, then $A/\mathcal{I}$ is the zero Cartier divisor. We write $0 = \operatorname{div}_C(A/A)$.
An effective Cartier divisor is a positive Cartier divisor or zero.

Consider an effective principal Cartier divisor $\operatorname{div}_C(x)$. Then $x^{-1}A$ is a free submodule of A and $\operatorname{div}_C(A/x^{-1}A)$ is an effective Cartier divisor. All free submodules of A are thus obtained and we know furthermore by Proposition 16.3, that $\operatorname{div}_C(x)$ and $\operatorname{div}_C(A/x^{-1}A)$ characterize each other. In other words, each effective principal Cartier divisor $\operatorname{div}_C(x)$ is an effective Cartier divisor (thank God!). Note by the way that $0 = \operatorname{div}_C(A/A) = \operatorname{div}_C(1)$.

Let $D = \operatorname{div}_C(A/\mathcal{I})$ and $D' = \operatorname{div}_C(A/\mathcal{J})$ be effective Cartier divisors of A. Note that $\mathcal{I}\mathcal{J}$ is a locally free submodule of A. Indeed if $\mathcal{I}A_P = aA_P$ and $\mathcal{J}A_P = bA_P$, with $a, b \in A_P$ regular elements, then $(\mathcal{I}\mathcal{J})A_P = \mathcal{I}A_P\mathcal{J}A_P = abA_P$. We put

$$D + D' = \operatorname{div}_C(A/\mathcal{I}\mathcal{J}).$$

Consequently, the set of effective Cartier divisors of A is equipped with an addition. Clearly, $0 = \operatorname{div}_C(1)$ is a neutral element for this addition (how surprising!).

We recall that the principal Cartier divisors form a commutative group and that by Proposition 16.5, the set of effective principal Cartier divisors is stable for the addition. Everything would fit nicely if this addition was induced by the operation on all effective Cartier divisors that we have just defined. This is indeed the case. More precisely, if $D = \operatorname{div}_C(A/x^{-1}A) = \operatorname{div}_C(x)$ and $D' = \operatorname{div}_C(A/y^{-1}A) = \operatorname{div}_C(y)$ are both principal effective Cartier divisors, we have

$$\begin{aligned} D + D' &= \operatorname{div}_C(A/x^{-1}y^{-1}A) = \operatorname{div}_C(A/(xy)^{-1}A) \\ &= \operatorname{div}_C(xy) = \operatorname{div}_C(x) + \operatorname{div}_C(y). \end{aligned}$$

We can now define the group of Cartier divisors of which the group of principal Cartier divisors is obviously a subgroup.

Definition 16.8 *Let A be a Noetherian ring.*

(i) *The group $\operatorname{Div}_C(A)$ of Cartier divisors of A is the commutative group generated by the set of effective Cartier divisors equipped with this addition.*

(ii) *Two Cartier divisors D and D' are said to be linearly equivalent if $D - D'$ is a principal Cartier divisor.*

(iii) *The quotient group of $\operatorname{Div}_C(A)$ by the group of principal Cartier divisors of A is the group $\operatorname{Cl}_C(A)$ of classes of Cartier divisors.*

Exercises 16.9

1. When A is a Noetherian domain, check that the groups of Cartier divisors and of classes of Cartier divisors are indeed the same as in Exercise 16.6.

2. Show that if R is a Noetherian local ring, then $\operatorname{Cl}_C(R) = 0$, i.e. all Cartier divisors of R are principal.

3. Show that if a Noetherian ring A has only finitely many maximal ideals, then $\operatorname{Cl}_C(A) = 0$.

Proposition 16.10 *Let A be a Noetherian ring and S be a multiplicatively closed part of A.*

If $A/\mathcal{I}$ is an effective Cartier divisor of A, then $S^{-1}A/\mathcal{I}S^{-1}A$ is an effective Cartier divisor of $S^{-1}A$. This induces group homomorphims

$$\mathrm{Div}_C(A) \to \mathrm{Div}_C(S^{-1}A) \quad and \quad \mathrm{Cl}_C(A) \to \mathrm{Cl}_C(S^{-1}A).$$

Proof Since $\mathcal{I}$ is a locally free ideal, $\mathcal{I}S^{-1}A$ is a locally free ideal of $S^{-1}A$. Furthermore, if $\mathcal{J}$ is another locally free ideal, we have

$$(\mathcal{I}\mathcal{J})S^{-1}A = (\mathcal{I}S^{-1}A)(\mathcal{J}S^{-1}A).$$

Finally, if $\mathrm{div}_C(A/\mathcal{I})$ is a principal effective Cartier divisor of A, then $\mathcal{I}$ is free, hence $\mathcal{I}S^{-1}A$ is free and $\mathrm{div}_C(S^{-1}A/\mathcal{I}S^{-1}A)$ is a principal effective Cartier divisor of $S^{-1}A$. $\qquad\square$

Definition 16.11 *If $D = \mathrm{div}_C(A/\mathcal{I})$ is an effective Cartier divisor of A, we define*
$$\mathrm{Supp}(D) = \mathrm{Supp}(A/\mathcal{I}).$$

Note that $\mathrm{Supp}(D+D') = \mathrm{Supp}(D) \cup \mathrm{Supp}(D')$. Indeed if $D = \mathrm{div}_C(A/\mathcal{I})$ and $D' = \mathrm{div}_C(A/\mathcal{J})$, it is clear that

$$\mathcal{I}\mathcal{J} \subset \mathcal{P} \Longleftrightarrow \mathcal{I} \subset \mathcal{P} \quad \text{or} \quad \mathcal{J} \subset \mathcal{P}.$$

Lemma 16.12

(i) *If $D = \mathrm{div}_C(A/\mathcal{I})$ is a positive Cartier divisor, all minimal prime ideals of $\mathcal{I}$ have height one.*

(ii) *The Support of an effective Cartier divisor contains only a finite number of height-one prime ideals.*

Proof (i) Let $\mathcal{P}$ be a minimal prime ideal of $\mathcal{I}$. Then $\mathcal{P}A_{\mathcal{P}}$ is a minimal prime ideal of $\mathcal{I}A_{\mathcal{P}}$. Since this last ideal is generated by a regular element, $\mathcal{P}A_{\mathcal{P}}$ has height one and $\mathcal{P}$ also.

If $D = \mathrm{div}_C(A/\mathcal{I})$, the prime ideals of height one containing $\mathcal{I}$ are the minimal prime ideals of $\mathcal{I}$, hence they are in finite number.

Consider now an integrally closed Noetherian domain R. Denote by K the fraction field of R. Let $\mathcal{P}$ be an irreducible Weil divisor of R and $v_{\mathcal{P}}$ the discrete valuation associated to it. If $D = \mathrm{div}_C(R/\mathcal{I})$ is an effective Cartier divisor, there exists $a \in K$ such that $\mathcal{I}_P = aR_P$. We note that $v_P(a)$ only depends on D and we put $v_P(D) = v_P(a)$. Clearly,

$$v_P(D) \neq 0 \Longleftrightarrow P \in \mathrm{Supp}(D) \Longleftrightarrow v_P(D) > 0,$$

hence this occurs only for a finite number of irreducible Weil divisors P. Furthermore, if $D' = \mathrm{div}_C(R/\mathcal{J})$ is another effective Cartier divisor, we have $v_P(D + D') = v_P(D) + v_P(D')$. We put $v_P(D - D') = v_P(D) - v_P(D')$. The map

$$v : \mathrm{Div}_C(R) \to \mathrm{Div}_W(R), \quad \Delta \to v(\Delta) = \sum_{\mathrm{ht}(P)=1} v_P(\Delta)[P]$$

is a well defined group homomorphism. $\square$

Theorem 16.13 *Let R be an integrally closed Noetherian domain.*

(i) *The map*

$$v : \mathrm{Div}_C(R) \to \mathrm{Div}_W(R), \quad \Delta \to v(\Delta) = \sum_{\mathrm{ht}(P)=1} v_P(\Delta)[P]$$

   is an injective group homomorphism.

(ii) *The Cartier divisor Δ is effective if and only if the Weil divisor $v(\Delta)$ is effective.*

(iii) *For each $x \in K(R)^*$, we have $v(\mathrm{div}_C(x)) = \mathrm{div}_W(x)$ and v induces therefore an injective group homomorphism $\bar{v} : \mathrm{Cl}_C(R) \to \mathrm{Cl}_W(R)$.*

(iv) *The homomorphism v is surjective if and only if R_M is a UFD for all maximal ideals M of R.*

Proof (i) We show that v is injective. Indeed, assume that $D = \mathrm{div}_C(R/\mathcal{I})$ and $D' = \mathrm{div}_C(R/\mathcal{J})$ are effective Cartier divisors such that $v_P(D) = v_P(D')$ for all irreducible Weil divisors P. This means $\mathcal{I}_P = \mathcal{J}_P$ for all height-one prime ideals P. Since a finitely generated locally free module is reflexive, by Corollary 15.12 we find

$$\mathcal{I} = \bigcap_{\mathrm{ht}(P)=1} \mathcal{I}_P = \bigcap_{\mathrm{ht}(P)=1} \mathcal{J}_P = \mathcal{J}.$$

Now, consider Cartier divisors Δ and Δ_1 such that $v(\Delta) = v(\Delta_1)$. There exist effective Cartier divisors D, D', D_1 and D'_1 such that $\Delta = D - D'$ and $\Delta_1 = D_1 - D'_1$. We find $v(D + D'_1) = v(D_1 + D')$, hence $D + D'_1 = D_1 + D'$ and $\Delta = \Delta_1$.

   (ii) If $D = \mathrm{div}_C(R/\mathcal{I})$ is an effective Cartier divisor, then $v_P(D) \geq 0$ for each irreducible Weil divisor P and $v(D)$ is effective. Conversely, put

$D' = \operatorname{div}_C(R/\mathcal{J})$ and assume that $v(D - D')$ is an effective Weil divisor. Let $\mathcal{P}$ be an irreducible Weil divisor. Since $v_{\mathcal{P}}(D) \geq v_{\mathcal{P}}(D')$, we have

$$(\mathcal{J} : \mathcal{I})_{\mathcal{P}} = \mathcal{J}_{\mathcal{P}} : \mathcal{I}_{\mathcal{P}} = \mathcal{P}^{v_{\mathcal{P}}(D')} R_{\mathcal{P}} : \mathcal{P}^{v_{\mathcal{P}}(D)} R_{\mathcal{P}} = \mathcal{P}^{v_{\mathcal{P}}(D-D')} R_{\mathcal{P}} \subset R_{\mathcal{P}}.$$

Hence $\mathcal{J} : \mathcal{I}$ is a locally free ideal and the effective Cartier divisor $\Delta = \operatorname{div}_C(R/(\mathcal{J} : \mathcal{I}))$ satisfies

$$v(\Delta) = v(D - D').$$

(iii) We deduce $v(\operatorname{div}_C(x)) = \operatorname{div}_W(x)$ from the definition of v. Consequently v induces an injective homomorphism $\mathrm{Cl}_C(R) \to \mathrm{Cl}_W(R)$.

(iv) Clearly, our homomorphism v is surjective if and only if all irreducible Weil divisors are Cartier divisors, in other words if and only if $\mathcal{P}$ is a locally free ideal for $h(\mathcal{P}) = 1$.

First we assume that $A_{\mathcal{M}}$ is a UFD for all maximal ideals $\mathcal{M}$. Then $A_{\mathcal{N}}$ is a UFD for all prime ideals $\mathcal{N}$. Let $\mathcal{P}$ be an irreducible Weil divisor. If $\mathcal{P} \subset \mathcal{N}$, then $\mathcal{P} A_{\mathcal{N}}$ is a prime ideal of height-one, hence a free $A_{\mathcal{N}}$-module of rank one. If $\mathcal{P} \not\subset \mathcal{N}$, then $\mathcal{P} A_{\mathcal{N}} = A_{\mathcal{N}}$ is also a free $A_{\mathcal{N}}$-module of rank-one. In other words $\mathcal{P}$ is a Cartier divisor.

Conversely, let $\mathcal{M}$ be a maximal ideal of A. Let $\mathcal{Q}$ be an irreducible divisor of $A_{\mathcal{M}}$. Then $\mathcal{P} = \mathcal{Q} \cap R$ is a height-one prime ideal of R. If $\mathcal{P}$ is locally free, then $\mathcal{Q} = \mathcal{P} A_{\mathcal{M}}$ is free, hence principal and generated by a regular element. We are done by Theorem 15.6. $\qquad\square$

Exercise 16.14 Let A be a Noetherian domain and K its fraction field.

1. Show that if $\mathcal{F} \subset K$ is a Cartier divisor of A, then $\mathcal{F}A[X] \subset K(X)$ is a Cartier divisor of $A[X]$.

2. Show that the map $\mathcal{F} \to \mathcal{F}A[X]$ from $\mathrm{Div}_C(A)$ to $\mathrm{Div}_C(A[X])$ is a group homomorphism.

3. Show that, when A is integrally closed, this group homomorphism induces an isomorphism $\mathrm{Cl}_C(A) \simeq \mathrm{Cl}_C(A[X])$.

16.2 The Picard group

In this section A is a given Noetherian ring. A finitely generated locally free A-module of rank one is called an invertible A-module.

Proposition 16.15

(i) *If L and L' are invertible A-modules, so is $L \otimes_A L'$.*

(ii) *This operation gives to the set of isomorphism classes of invertible A-modules the structure of a commutative group.*

Proof (i) Let $\mathcal{P}$ be a prime ideal of A. By Proposition 7.21, we have $(L \otimes_A L')_{\mathcal{P}} \simeq L_{\mathcal{P}} \otimes_{A_{\mathcal{P}}} L'_{\mathcal{P}}$. Since $L_{\mathcal{P}} \simeq A_{\mathcal{P}} \simeq L'_{\mathcal{P}}$, we are done.

(ii) The natural isomorphisms

$$(L \otimes_A L') \otimes_A L'' \simeq L \otimes_A (L' \otimes_R L''), \quad (L \otimes_A L') \simeq (L' \otimes_A L) \quad \text{and} \quad A \otimes_A L \simeq L$$

show immediately that the operation is associative and commutative and that the isomorphism class of A is a neutral element. We are left with finding an inverse for each element. We claim that the finitely generated rank one, locally free module $L^{\check{}}$ is the inverse of L, in other words that $(L^{\check{}} \otimes_A L) \simeq A$. Since the homomorphism

$$(L^{\check{}} \otimes_A L) \to A, \text{ with } f \otimes_A x \to f(x)$$

is obviously an isomorphism, this is clear.   □

Definition 16.16 *The group of isomorphism classes of invertible A-modules is the Picard group* $\mathrm{Pic}\,(A)$ *of A.*

If L is an invertible A-module, we denote the class of L in $\mathrm{Pic}\,(A)$ *by $[L]$ and we put $[L][L'] = [L \otimes_A L']$.*

Note, it is clear, that when R is an integrally closed domain, then the Picard group $\mathrm{Pic}\,(R)$ is a subgroup of the group of isomorphism classes of quasi-invertible A-modules.

Theorem 16.17 *There is a natural surjective group homomorphism*

$$\pi_C : \mathrm{Div}\,_C(A) \to \mathrm{Pic}\,(A)$$

whose kernel is the group of principal divisors and inducing an isomorphism $\mathrm{Cl}_C(A) \simeq \mathrm{Pic}\,(A)$.

Proof For each effective Cartier divisor $D = \mathrm{div}\,_C(A/\mathcal{I})$, we define $\pi_C(D) = [\mathcal{I}^{\check{}}]$. Consider $D' = \mathrm{div}\,_C(A/\mathcal{J})$. Then

$$\pi_C(D + D') = [(\mathcal{I}\mathcal{J})^{\check{}}] \simeq [\mathcal{I}^{\check{}} \otimes_A \mathcal{J}^{\check{}}] = [\mathcal{I}^{\check{}}][\mathcal{J}^{\check{}}].$$

Hence π is a map from the set of effective Cartier divisors to $\mathrm{Pic}\,(A)$ which commutes with the operations in $\mathrm{div}\,_C(A)$ and $\mathrm{Pic}\,(A)$. Since the group $\mathrm{div}\,_C(A)$ is generated by the effective Cartier divisors, we have defined our homomorphism.

Now we show that π is surjective. Consider $[L] \in \mathrm{Pic}\,(A)$. By Proposition 7.70, there exists an injective homomorphism $f : A \to L$ inducing an isomorphism

$$f_{\mathcal{P}} : A_{\mathcal{P}} \to L_{\mathcal{P}} \text{ for } \mathcal{P} \in \mathrm{Ass}(A) = \mathrm{Ass}(L).$$

We claim that the transposed homomorphism $f^\vee : L^\vee \to A^\vee \simeq A$ is injective. Let K be its kernel. Since $f_{\mathcal{P}}^\vee$ is an isomorphism for all $\mathcal{P} \in \text{Ass}(A) = \text{Ass}(L^\vee)$, we have $K_{\mathcal{P}} = (0)$ for all $\mathcal{P} \in \text{Ass}(K) \subset \text{Ass}(L^\vee)$, hence $K = (0)$. Consequently, $L^\vee$ is isomorphic to an ideal $\mathcal{I} \subset A$. Since L is invertible, so is $\mathcal{I}$. Then $D = \text{div}_C(A/\mathcal{I})$ is an effective Cartier divisor such that $L \simeq \mathcal{I}^\vee$, i.e. $[L] = \pi_C(D)$.

Next we show that the kernel of π_C is the group of principal Cartier divisors. Consider a principal Cartier divisor $\text{div}_C(ab^{-1})$, where $a, b \in A$ are regular elements. We have

$$
\begin{aligned}
\pi_C(\text{div}_C(ab^{-1})) &= \pi_C(\text{div}_C(b^{-1}))\pi(\text{div}_C(a^{-1}))^{-1} \\
&= \pi_C(\text{div}_C(A/bA))\pi_C(\text{div}_C(A/aA))^{-1} \\
&= [Ab^{-1}][Aa^{-1}]^{-1} = [Ab^{-1} \otimes_A Aa] \simeq [A].
\end{aligned}
$$

Conversely, if $D = \text{div}_C(A/\mathcal{I})$ and $D' = \text{div}_C(A/\mathcal{J})$ are effective Cartier divisors such that $\pi_C(D - D') = [A]$, we have

$$ \mathcal{I}^\vee \otimes_A \mathcal{J} \simeq A. $$

We want to show that there exists a regular element x in the total quotient ring $T(A)$ such that $\text{div}_C(x) = D - D'$, in other words such that $x\mathcal{I} = \mathcal{J}$. By Proposition 7.70, there exists an injective homomorphism $i : A \to \mathcal{I}$. Put $a = i(1)$. Since $xa = xi(1) = i(x)$ for all $x \in A$, it is clear that $a \in \mathcal{I}$ is a regular element. Consequently $a\mathcal{J}$ is a Cartier divisor contained in $\mathcal{I}$. We claim that there exists a regular element $b \in A$ such that $b\mathcal{I} = a\mathcal{J}$. First we show that the natural homomorphisms

$$ h : (a\mathcal{J} : \mathcal{I}) \to \text{Hom}_A(\mathcal{I}, a\mathcal{J}) \quad \text{and} \quad j : \mathcal{I}^\vee \otimes_A a\mathcal{J} \to \text{Hom}_A(\mathcal{I}, a\mathcal{J}) $$

are isomorphisms. If $\mathcal{P}$ is a prime ideal of A, there exist regular elements $x, y \in A_{\mathcal{P}}$ such that $\mathcal{I}_{\mathcal{P}} = xA_{\mathcal{P}}$ and $\mathcal{J}_{\mathcal{P}} = yA_{\mathcal{P}}$. Clearly,

$$ h_{\mathcal{P}} : (a\mathcal{J}_{\mathcal{P}} : \mathcal{I}_{\mathcal{P}}) = (ayA_{\mathcal{P}} : xA_{\mathcal{P}}) \to \text{Hom}_{A_{\mathcal{P}}}(xA_{\mathcal{P}}, ayA_{\mathcal{P}}) \quad \text{and} $$

$$ j_{\mathcal{P}} : (xA_{\mathcal{P}}^\vee \otimes_{A_{\mathcal{P}}} ayA_{\mathcal{P}}) \to \text{Hom}_{A_{\mathcal{P}}}(xA_{\mathcal{P}}, ayA_{\mathcal{P}}) $$

are isomorphisms, hence h and j are isomorphisms by Corollary 7.28. Consequently, we find

$$ a\mathcal{J} : \mathcal{I} \simeq \mathcal{I}^\vee \otimes_A a\mathcal{J} \simeq \mathcal{I}^\vee \otimes_A \mathcal{J} \simeq A $$

and the ideal $a\mathcal{J} : \mathcal{I}$ is free. Let $b \in A$ be a regular element such that $a\mathcal{J} : \mathcal{I} = bA$. This implies $b\mathcal{I} \subset a\mathcal{J}$. Since $a\mathcal{J}_{\mathcal{P}}$ and $\mathcal{I}_{\mathcal{P}}$ are free for each prime ideal $\mathcal{P}$, this shows furthermore $b\mathcal{I}_{\mathcal{P}} = a\mathcal{J}_{\mathcal{P}}$ for each prime ideal $\mathcal{P}$, hence $b\mathcal{I} = a\mathcal{J}$ by Corollary 7.28. $\qquad\square$

Theorem 16.18 *If R is integrally closed, the homomorphism $\pi_C : \operatorname{Div}_C(R) \to$ $\operatorname{Pic}(R)$ is the restriction to $\operatorname{Div}_C(R)$ of the natural homomorphism π_W from $\operatorname{Div}_W(R)$ to the group of isomorphism classes of quasi-invertible R-modules.*

Proof We recall that if $D = \operatorname{div}_W(R/\mathcal{I})$ is an effective Weil divisor, we defined $\pi_W(D) = [R :_K \mathcal{I}] = [\mathcal{I}^-]$. If furthermore D is a Cartier divisor, i.e. if $\mathcal{I}$ is invertible, then we defined $\pi_C(D) = [\mathcal{I}^-] \in \operatorname{Pic}(R)$, so the theorem is clear.

$\square$

Exercise 16.19 Let A be a Noetherian ring and B a Noetherian A-algebra.

1. Show that if L is an invertible A-module, $L \otimes_A B$ is an invertible B-module.

2. Show that the map $\operatorname{Pic}(A) \to \operatorname{Pic}(B)$ thus defined, is a group homomorphism.

3. When $B = A[X]$ and A is integrally closed, show that this homomorphism is an isomorphism.

16.3 Exercises

1. Let $D_1, \ldots, D_n$ be positive Cartier divisors in a Noetherian ring A. Show that for each minimal prime $\mathcal{P}$ of the closed set $\bigcap_{i=1}^n \operatorname{Supp}(D_i)$, we have $\operatorname{ht}(\mathcal{P}) \leq n$.

   Definition 16.20 *If for each minimal prime ideal $\mathcal{P}$ of $\bigcap_{i=1}^n \operatorname{Supp}(D_i)$, we have $\operatorname{ht}(\mathcal{P}) = n$, then we say that the positive divisors D_i intersect properly.*

2. Let A be a Noetherian ring and let $D = \operatorname{div}_C(A/\mathcal{I})$ and $D' = \operatorname{div}_C(A/\mathcal{J})$ be two positive Cartier divisors intersecting properly. Show that $(\mathcal{I} + \mathcal{J})/\mathcal{I} \subset A/\mathcal{I}$ is a locally free ideal if and only if $\operatorname{Supp}(A/\mathcal{J}) \cap \operatorname{Ass}(A/\mathcal{I}) = \emptyset$.

3. Let A be a Noetherian ring and $D = \operatorname{div}_C(A/\mathcal{I})$ and $D' = \operatorname{div}_C(A/\mathcal{J})$ two positive Cartier divisors. Show that there exists an effective Cartier divisor $\Delta = \operatorname{div}_C(A/\mathcal{J}')$ linearly equivalent to D' and such that $(\mathcal{I} + \mathcal{J}')/\mathcal{I} \subset A/\mathcal{I}$ is a locally free ideal.

4. Let R be an integrally closed Noetherian ring. Consider a Weil divisor $\operatorname{div}_W(R/\mathcal{I})$ and a Cartier divisor $\operatorname{div}_C(R/\mathcal{J})$. Assume that $R/\mathcal{I}$ and $R/\mathcal{J}$ have no common component, in other words that $R/(\mathcal{I} + \mathcal{J})$ is of codimension at least two. Show that $(\mathcal{I} + \mathcal{J})/\mathcal{I}$ is a locally free submodule of the ring $R/\mathcal{I}$, hence that $R/(\mathcal{I} + \mathcal{J})$ defines an effective Cartier divisor of the ring $R/\mathcal{I}$.

5. Let A be a Noetherian ring and $a \in A$. Assume that $a \in JR(A)$. Show that the natural group homomorphism $\text{Pic}\,(A) \to \text{Pic}\,(A/aA)$, $[L] \to [L/aL]$, is injective.

6. Let A be a Noetherian ring and $a \in A$ a regular element such that aA is a prime ideal. Show that the natural homomorphism $\text{Pic}\,(A) \to \text{Pic}\,(A_a)$ defined by $[L] \to [L_a]$, is an isomorphism.

7. Let A be a Noetherian ring and $a \in A$ a regular element. Assume that A_a is a UFD. Show that $\text{Pic}\,(A)$ is a finitely generated group.

8. Consider a Noetherian ring A, a non-invertible element $a \in A$ and the natural group homomorphism $\rho : \text{Pic}\,(A) \to \text{Pic}\,(A/aA)$. Let L be an invertible A-module. Show that $[L] \in \ker \rho$ if and only if there exists an effective Cartier divisor $D = \text{div}_C(A/\mathcal{I})$ such $[L] = [\mathcal{I}^-]$ and that $\text{Supp}(D) \cap V(aA) = \emptyset$.

Bibliography

[1]   M.F. Atiyah and I.G. Macdonald, *Introduction to Commutative Algebra*, Addison-Wesley Series in mathematics.

[2]   N. Bourbaki, *Algèbre*, Hermann, Paris.

[3]   N. Bourbaki, *Algèbre Commutative*, Hermann, Paris.

[4]   D. Eisenbud, *Commutative Algebra with a View toward Algebraic Geometry*, Graduate Texts in Mathematics 150, Springer Verlag.

[5]   A. Grothendieck and J. Dieudonné, *Eléménts de Géometrie Algébrique*, Publications mathématiques de l'I.H.E.S.

[6]   W. Krull, *Idealtheory*, Ergebnisse der Mathematik, Springer Verlag.

[7]   M.P. Malliavin, *Algèbre Commutative*, Masson.

[8]   H. Matsumura, *Commutative Algebra*, W.A. Benjamin.

[9]   H. Matsumura, *Commutative Ring Theory*, Cambridge Studies in Advanced Mathematics 8, Cambridge University Press

[10] M. Nagata, *Local Rings*, Interscience publishers.

[11] D.G. Northcott, *Ideal Theory*, Cambridge University Press.

[12] D.G. Northcott, *Lessons on Rings, Modules and Multiplicities*, Cambridge University Press.

[13] J.P. Serre, *Algèbre Locale. Multiplicités*, Lecture Notes in Mathematics 11, Springer Verlag.

[14] R.Y. Sharp, *Steps in Commutative Algebra*, London Mathematical Society Students Texts 19, Cambridge University Press.

[15] B.L. Van der Waerden, *Modern Algebra*, Springer Verlag.

[16] O. Zariski and P. Samuel, *Commutative Algebra*, Van Nostrand, Princeton.

Subject index

Symbols index